THE
VISIONEERS

THE
VISIONEERS

How a Group of Elite
Scientists Pursued Space
Colonies, Nanotechnologies,
and a Limitless Future

W. Patrick McCray

Princeton University Press

Princeton and Oxford

Copyright © 2013 by Princeton University Press
Published by Princeton University Press, 41 William Street, Princeton,
New Jersey 08540
In the United Kingdom: Princeton University Press, 6 Oxford Street,
Woodstock, Oxfordshire OX20 1TW

press.princeton.edu

Library of Congress Cataloging-in-Publication Data

McCray, Patrick (W. Patrick)
  The visioneers : how a group of elite scientists pursued space
colonies, nanotechnologies, and a limitless future / W. Patrick McCray.
     p. cm.
  Summary: "In 1969, Princeton physicist Gerard O'Neill began looking outward to
space colonies as the new frontier for humanity's expansion. A decade later, Eric Drexler,
an MIT-trained engineer, turned his attention to the molecular world as the place where
society's future needs could be met using self-replicating nanoscale machines. These modern utopians predicted that their technologies could transform society as humans mastered
the ability to create new worlds, undertook atomic-scale engineering, and, if truly successful, overcame their own biological limits. The Visioneers tells the story of how these scientists and the communities they fostered imagined, designed, and popularized speculative
technologies such as space colonies and nanotechnologies. Patrick McCray traces how
these visioneers blended countercultural ideals with hard science, entrepreneurship, libertarianism, and unbridled optimism about the future. He shows how they built networks
that communicated their ideas to writers, politicians, and corporate leaders. But the visioneers were not immune to failure—or to the lures of profit, celebrity, and hype. O'Neill
and Drexler faced difficulty funding their work and overcoming colleagues' skepticism,
and saw their ideas co-opted and transformed by Timothy Leary, the scriptwriters of Star
Trek, and many others. Ultimately, both men struggled to overcome stigma and ostracism
as they tried to unshackle their visioneering from pejorative labels like "fringe" and "pseudoscience." The Visioneers provides a balanced look at the successes and pitfalls they encountered. The book exposes the dangers of promotion—oversimplification, misuse, and
misunderstanding—that can plague exploratory science. But above all, it highlights the
importance of radical new ideas that inspire us to support cutting-edge research into tomorrow's technologies"— Provided by publisher.
  Includes bibliographical references and index.
  ISBN 978-0-691-13983-8 (hardback : acid-free paper)   1. Science—History.
2. Visionaries.   I. Title.
  Q125.M417 2012
  509—dc23   2012017061

British Library Cataloging-in-Publication Data is available

This book has been composed in Sabon LT Std with Helvetica Neue display

Printed on acid-free paper. ∞

Printed in the United States of America

10 9 8 7 6 5 4 3 2 1

We who seek adventure everywhere
We are not your enemies
We want to give you vast and strange domains
Where flowering mystery offers itself to those who wish to pluck it . . .
Pity us who fight always at the frontiers
Of the limitless and the future
    —Guillaume Apollinaire, "La jolie rousse" (The Pretty Redhead),
        Calligrammes, 1918

Contents

||||||||||||||||||||||||||||||||||||||||||||||||||||||||||||||||||||||||||||||||||||||||||||||||||||||||||||||||||||||||||||||||||||||||||||

Illustrations

Acknowledgments

Many people, some of whom also appear as actors in this book, graciously lent me materials from their personal collections and took the time to answer my queries about matters both large and insignificant. While they may not agree with all of my conclusions or interpretations, the resources they shared helped build the foundation this book rests upon. In no especial order, my thanks to Dale Amon, Trudy E. Bell, Ben Bova, Stewart Brand, Taylor Dark III, K. Eric Drexler, Freeman Dyson, Don Eigler, Robin Hanson, Keith Henson, Mark Hopkins, Tom Kalil, Henry Kolm, Ralph Merkle, Mark S. Miller, Christine L. Peterson, Gayle Pergamit, Rusty Schweickart, and Ned Seeman. David Brandt-Erichsen, James C. Bennett, and Conrad Schneiker deserve special recognition for generously sharing extensive materials from their personal collections. Tasha O'Neill gave me free rein of her basement in Princeton, New Jersey, to explore the personal papers of her late husband, Gerard O'Neill.

Several students at the University of California, Santa Barbara—Mary-Ingram Waters, Roger Eardley-Pryor, Samantha Rohman, Olivia Russell, and Sabrina Wuu—tracked down hard-to-find research materials.

Funding for writing this book came from a number of sources. Foremost among these is the Center for Nanotechnology in Society at the University of California, Santa Barbara, which the National Science Foundation funded under Grant No. SES 0531184. The opinions, findings, and conclusions or recommendations expressed here are mine alone and don't necessarily reflect views of the NSF. I began writing this book in 2008 while in Cassis, France, via a fellowship from the Camargo Foundation. I finished it in

2012 as a visiting professor at the California Institute of Technology. In between these two fortunate opportunities, a Collaborative Research Fellowship from the American Council of Learned Societies provided additional time to write.

Many colleagues and friends provided insights, critiques, and inspiration. Credit for where this book succeeds must be shared with them while demerits for not always heeding advice offered remain with me. I'd especially like to acknowledge Peter Alagona, Sarah Bagby, Michael Bess, David Brock, Glenn Bugos, Hyungsub Choi, Angela Creager, Luis Campos, Matt Eisler, Brice Erickson, Fiona Goodchild, Evelyn Hu, Ann Johnson, David Kaiser, Bruce Lewenstein, Henry Lowood, Mara Mills, Joe November, Kathy Olesko, Nick Rasmussen, Luke Roberts, Paul Saffo, Josh Schimel, Howard Segal, Chris Toumey, Fred Turner, Spencer Weart, Peter Westwick, and Matt Wisnioski. Especial gratitude goes to my colleagues Michael Gordin and Cyrus Mody, who took time from their own writing to provide generous suggestions and gentle critiques. Finally, for those times when this book's future seemed less clear, I want to thank Nicole Archambeau.

W. Patrick McCray
June 2012, Pasadena, California

THE
VISIONEERS .
|||||||||||||||||||||||||||||||

Visioneering Technological Futures

I am vitally interested in the future, because I am going to spend the
rest of my life there.
 —Charles F. Kettering, inventor and head of General Motors' corporate
  research, quoted in advertising campaign for *Omni* magazine, 1978

On August 11, 1977, some 1,100 invited guests trekked to the old
Museum of Science and Industry in downtown Los Angeles and
celebrated California's first Space Day. Space exploration was big
news that summer. At theaters all across the United States, *Star
Wars* was raking in millions of dollars as fans queued to see the
epic space opera over and over. The upsurge of excitement about
space wasn't limited to just the silver screen. Out in California's
Mojave Desert, engineers were readying the space shuttle *Enterprise* for its first solo atmospheric flight. NASA had high hopes
that America's human spaceflight program, stagnant since the end
of the Apollo era, would be revived by its new "space truck."

Buttoned-down aerospace executives, anxious NASA managers,
and cynical politicos mingled with celebrity scientists like astronomer Carl Sagan and oceanographer Jacques Cousteau in the museum's elegant rose-festooned garden. Sprinkled among the guests
were real astronauts who had been to space and *Star Trek* fans
who yearned to go. Counterculture icons like LSD guru Timothy
Leary and *Whole Earth Catalog* publisher Stewart Brand enliv-
ened the crowd, which swelled even more when California governor Edmund "Jerry" Brown and his entourage arrived.

Accompanied by a phalanx of journalists, the invitees moved
into the museum's Progress Hall and took seats beneath suspended
satellites and rockets left over from the triumphant days of the
space race. But the audience had little interest in revisiting what

the United States had already accomplished in space. They wanted to hear about the future. And, for quite a few in the crowd, the future meant space colonies.

Princeton physicist Gerard O'Neill—tall and trim with a Beatles-style haircut and a preference for turtleneck sweaters underneath his favorite Harris tweed sports jacket—was a prime catalyst for the event. For years, O'Neill had designed and promoted the construction of massive earthlike habitats that would float free in outer space far from our home planet's gravitational pull. He began with simple drawings and back-of-the-envelope estimates. In time, O'Neill's concepts matured into sophisticated designs backed by detailed calculations that he disseminated and discussed with others who shared his passion. Sensing a rise in public interest, O'Neill was ready to put his plans into action.

Time, the physicist believed, was of the essence. O'Neill imagined that the "humanization of space" could provide a respite for a crowded, polluted, and energy-hungry planet.[1] In the future, space could be not just a government-run *program* for astronaut elites but a *place* where ordinary citizens could live and work. In early 1977, O'Neill gathered his plans and hopes for space colonization into an award-winning book called *The High Frontier*, and its publication ignited more interest in his ideas.[2]

A wildly eclectic assortment of speakers, diverse in their occupations as well as their opinions about humanity's future in space, took the podium at Space Day. Beat generation writer Michael Mc-Clure read a poem that reflected his hopes that ecology and technology could find a sense of unity in space. "Join me here," he said, "in this space that we invent from real stuff where we have never laughed, nor danced, nor sung before."[3] Other speakers echoed environmental themes by describing future satellites that could collect and beam down clean energy produced by efficiently collecting the boundless solar power found in space. Outside the hall, meanwhile, Timothy Leary pushed his psychedelic version of O'Neill's vision. "Now there is nowhere left for smart Americans to go but out into high orbit. I love that phrase—high orbit," Leary vamped to reporters. "We were talking about high orbit long before the space program."[4]

When it was O'Neill's turn to address the crowd, he refuted Carl Sagan's suggestion that future space exploration would best be left to robots and machines. Taking a cue from the summer's most popular movie, he announced that "it was time to stop R2-D2 from having all the fun" and critiqued America's current "timid" space ambitions. Instead, O'Neill championed a strategy of "direct human involvement," which he illustrated with dazzling images of life aboard space-based settlements of the future.[5]

But O'Neill could show off more than just inspiring pictures. He also displayed a working prototype of a "mass driver" that he had helped design and build. The device used electromagnetic fields to accelerate payloads to very high speeds. O'Neill envisioned a much larger machine—something students at the Massachusetts Institute of Technology were already working on—that could propel minerals from the moon or an asteroid out into space where they would be recovered and processed into materials for building settlements in space. Such would be the trees that settlers of the future would use to build the log cabins of the new frontier O'Neill imagined.

George Koopman, an independently wealthy Los Angeleno and occasional host of a New Age–inflected radio show about radical technologies, listened to the talks at Space Day. Koopman was a space enthusiast who, years later, would start a company that set out to make rockets. But, in 1977, Koopman was gathering information for the L5 Society, a grassroots pro-space group enthusiasts formed in response to O'Neill's vision. For Koopman, there was only one way to interpret what he heard: "We're going."[6]

Not everyone was as convinced. Underground artist R. Crumb, creator of '70s-era comics such as *Keep on Truckin'* and *Fritz the Cat*, was in the audience too. Space Day ("or whatever the hell it was called," he wrote) infuriated Crumb. The museum was just "a show room for aero-space corporations," while O'Neill and other space enthusiasts were a "smug bunch of hypocrits [*sic*]." That so many people were "falling for the space hype hook, line, and sinker" finally prompted Crumb to stalk out in disgust.[7]

While Crumb fumed outside the museum, he could have encountered laid-off workers waving signs proclaiming "Jobs on Earth, Not in Space" and "Brown, Hire an Earthling" that expressed more terrestrial concerns. Within Jerry Brown's own cabinet, where support for "soft technologies" like solar power and wind energy ran high, one gubernatorial adviser grumbled, "This is disgusting. It's a technology worship session."[8] O'Neill's expansive vision for the humanization of space, which he believed offered profound environmental benefits, clearly had the power to infuriate as well as inspire.

When Jerry Brown spoke, he stood in front of a giant photo of the earth from space (courtesy of NASA) emblazoned with Space Day's motto: "California in the Space Age: An Era of Possibilities." The future Brown described, with energy beamed to earth from orbiting satellites and space settlements providing a "safety valve of unexplored frontiers," was drawn straight from O'Neill's vision.[9] Reflecting on the banner behind him, Brown said, "It is a world of limits but through respecting and reverencing the limits, endless possibilities emerge. . . . As for space colonies, it's not a question of whether—only when and how."[10]

It was no accident that Brown used the word "limits." A national magazine titled its 1975 profile of Brown as "Learning to Live With Our Limits." A onetime Jesuit seminarian who professed a fondness for asceticism and Zen retreats, Brown often warned his constituents of new constraints unfamiliar to those accustomed to the Golden State's seemingly boundless prosperity. "This country is entering an era of limits," he proclaimed in a national television address during the 1976 presidential campaign. "We're all on it, Spaceship Earth, hurtling through the universe."[11] A cartoon in a Sacramento newspaper captured this dismal view of the future with a road sign announcing: "Entering California. Lower Your Expectations."[12]

In fact, "limits" emerged in the 1970s as one of the decade's watchwords. For more than two centuries, Americans had propelled themselves and their country into a technological torrent with enthusiasm that bordered on faith. Technology—the assembly line, the Bomb, the freeway, the silicon chip—had enabled the United States to become an economic, military, and cultural power. But, in the late 1960s, many Americans had started to loudly and sometimes violently question technology's ability to resolve society's problems. Fears of environmental catastrophe and nuclear war coupled with anxieties about resource depletion and overpopulation had strained their optimism to the breaking point. By the time Nixon's presidency was embroiled in scandal, a new sense of the future, one constrained by limits and scarcity, had emerged.[13]

But Jerry Brown's use of the word "limits" in his speeches reflected more than an awareness that the political and economic power of California (and America) was not infinite. Brown's comments—indeed, the attitudes expressed by many at Space Day—were a direct response to an event that had happened five years earlier.

In the 1970s, reporters spilled much ink explaining the Club of Rome's origin and purpose. Descriptions ranged from a "loose aggregation" of elite jet-setters studying the "future of man and the earth" (albeit with "a certain smugness") to a group of "crackpots" who believed the planet's future depended on "a Copernican change of vision." In reality, its thirty-some members formed an "invisible college" of high-profile businessmen, politicians, and scientists. They held their first major meeting at the venerable Accademia dei Lincei in Rome, hence the group's name.[14]

In March 1972, the Club of Rome released an influential report called *The Limits to Growth*.[15] Announced with a media blitz aimed at policy makers and ambassadors, its "doomsday timetable" predicted an inevitable collapse of societies all around the planet unless politicians and business leaders had the courage to restrict the growth of populations, industrialization, and resource use.[16] Instead of continued expansion, it called for economic and ecological equilibrium commensurate with a species wholly dependent on limited planetary resources. Extensive computer-based calculations by researchers from MIT provided the Club of Rome with evidence needed to support its bleak assessment of the future.

Although many scientists and economists savaged the methodology that produced *Limits*, the Club of Rome's report sent a powerful message about the constraints on what technology could accomplish for the future. Released as a paperback book, *Limits* became a global sensation. Its troubling conclusions compelled more than eight million people to buy copies, and *Limits* was translated into some thirty languages. The ideas in *Limits* infiltrated popular culture as well, providing unsettling themes for movies, television shows, and fiction well into the Reagan era.

By the time Jerry Brown spoke at Space Day, "limits" was a term charged with conflicting meanings. For some, it meant austerity, self-denial, and living responsibly with a small planetary footprint. To others, limits meant narrowed options, restricted political freedoms, and vastly lower expectations as to what technology could offer Americans. To both groups, "limits" served as a shibboleth. Said in the right context and company, it told people which of these divergent views of the future you imagined.

There was a third point of view however. Some optimistic and entrepreneurial-minded scientists and engineers saw the notion of limits not as a warning or impediment but as a challenge. Was the future *really* going to be this dire? No, they said. Instead, they trusted that unexplored technological solutions could offer a reprieve or even an escape for the United States and perhaps the planet. However, the threat of economic, environmental, and planetary limits rumbling in the distance provided an essential foil against which to contrast their visions of a limitless future. As these scientists and engineers imagined it, the future would break sharply with the past when people mastered the ability to create new worlds and build new, powerful machines using nanotechnology. For them, the present, with its doomsayers agonizing over constraints, was merely a prototype, a provisional plan of what would become a magnificent and far less limited future.

This book explores how and why this select group of scientists and engineers developed their broad and expansive visions of how the future could be made radically different through as-yet-undeveloped technologies. It looks at how their visions for these technological futures were promoted, embraced, and rejected. Depictions of these futures and their enabling technologies fostered communities of enthusiasts who read about, debated, and helped publicize their ideas. Such imaginings of the future made possible by space colonization, for example, or radical nanotechnologies attracted attention from journalists, artists, business leaders, and politicians. These visions also took hold in popular culture and helped create a picture for the public of what the future might be like. By the book's end, we'll see how some aspects of these imagined futures happened, although almost always not as their advocates imagined.

This book focuses on two particular visions of the future and the ensembles of technologies seen as critical to achieving them. Both imagined futures were catalyzed by advocates' belief that new technologies offered radical solutions that could defuse the threat of limits. One of these is Gerard O'Neill's ideas for settlements and factories in space, technologies he saw as an alternative to terrestrial limits, lifestyles, and manufacturing.

Although O'Neill shared a vision of space settlements, right down to how they might look, with "blue-skying" futurists and science fiction writers, his work differed from theirs in several important ways. First, where earlier visionaries offered descriptive speculations, O'Neill deployed extensive mathematical calculations and careful but bold extrapolations of existing technological trends to develop rigorously detailed plans for space settlements. Throughout the 1970s, he continued to refine and improve his initial designs, taking into account new data he collected and critiques from colleagues. O'Neill supplemented his pen-and-paper work by tirelessly promoting his vision to colleagues, interested citizens, politicians, and journalists. O'Neill's program for the humanization of space sparked a small-scale social movement in the wake of the Apollo program's termination. College students and other members of the baby boom generation proved especially vigorous supporters of a reinvigorated space program in which they imagined they could play a role.

One of these young adults was K. Eric Drexler. As an MIT undergraduate in the early 1970s, Drexler was drawn to O'Neill's ideas and designs for the humanization of space. Drexler went on to develop plans for lunar factories, solar sails, and methods to mine asteroids for mineral resources. Besides patenting some of his ideas, he also helped build the mass-driver prototype displayed at

California's Space Day. By his early twenties, Drexler was one of the L5 Society's most articulate and vocal advocates for an expanded human presence in space.

Starting in the late 1970s, however, Drexler began to envision a new technological frontier. Whereas space promised the infinitely vast, nanotechnology shifted attention toward the molecular scale. (A nanometer, a basic unit of nanotechnology, is a mere one-billionth of a meter or about the size of a sugar molecule and far smaller than a virus. Put another way, a nanometer is to a meter roughly what a child's marble is to the size of the earth.) Inspired by steady advances that engineers and scientists had made in microelectronics and molecular biology, Drexler blended and extrapolated these in new directions to imagine a future in which people designed and built nanoscale materials, structures, and machines with near-atomic precision.

Through popular books, articles, and technical papers, Drexler and his supporters described self-replicating "universal assemblers" that might one day refashion the material world "from the bottom up, putting every atom in its place."[17] The power of nanotechnology, Drexler predicted in his influential 1986 book *Engines of Creation*, could mean more efficient use of natural resources, manufacturing that was less environmentally destructive, and even the ultimate set of tools for reaching beyond the earth into space.[18] The fact that he titled one book chapter "Limits to Growth" and another "Engines of Abundance" speaks to the enduring effects of the Club of Rome's report. Well into the 1990s, Drexler, aided by articles about his ideas in magazines such as *Omni*, the *Economist*, and even *Reader's Digest*, successfully promoted nanotechnology even as some researchers refuted claims as to what sorts of nanoscale machines one could possibly build.

Space colonization and nanotechnology. At first sight they present an odd combination. But this pairing of technologies makes sense. For much of the twentieth century, space exploration was *the* archetypal technological frontier, the blank slate on which generations of engineers and schoolchildren projected their wildest dreams. But techno-dreaming shifted to new realms. By the late 1990s, following Drexler's popularization, researchers, venture

Figure I.1 Prototype mass driver, c. 1977, built at MIT, with (*from right*) MIT student K. Eric Drexler, MIT professor Henry Kolm, Princeton professor Gerard O'Neill, and three other MIT students. (Image courtesy of Tasha O'Neill.)

capitalists, and policy makers were declaring nanotechnology as the critical new technological frontier for the twenty-first century. The most radical schemes, while often popular with general readers, were resisted and ridiculed as impracticable. Yet aspects of these futuristic nano visions were co-opted into more mainstream plans for research and technology development.

Despite their vastly different scales, futuristic concepts for settlements in space and for nanotechnology both centered on the mastery of the material world through technology. This book shows how the two topics, their promoters, and the communities that coalesced around them overlapped and proved influential in surprising and unexpected ways. In both cases, proponents imagined building a limitless tomorrow that sidestepped catastrophist scenarios of the future to offer endless space to expand, an abundance of resources, and, in the most radical versions, the possibility of transcending the mortal limits of the human body itself.

Visioneers

O'Neill's and Drexler's work pushed far beyond speculation. This alone distinguishes them from futurists content to prognosticate from a podium. O'Neill, Drexler, and others like them used their training in science and engineering to undertake detailed design and engineering studies. As a result, O'Neill and Drexler could explore and develop their conceptions of an expansive future created by the technologies they studied, designed, and promoted. In some cases, their studies led to the creation of actual things: prototypes, models, patents, and computer simulations. Just as importantly, O'Neill and Drexler also built communities and networks that connected their ideas to interested citizens, writers, politicians, and business leaders.

We lack an appropriate term for someone who undertook such a diverse set of future-directed activities. To fill this gap, I propose *visioneer*. A neologism combining "visionary" and "engineer," this word captures the hybridized nature of these technologists' activities.[19] By using O'Neill and Drexler as archetypal examples, this book explores the role of visioneers over the last forty years as they proposed and promoted new technologies.

The visionary aspect is essential to understanding visioneers' motivation. O'Neill, Drexler, and the communities they helped foster imagined that their technologies could shape future societies, upend traditional economic models, and radically transform the human condition. These plans flirted with and sometimes embraced technological utopianism. Imagining the ramifications of settlements and factories in space or nanoscale machines was more ambitious, yet far more challenging to realize, than designing a faster airplane or new computer circuit. But visioneers' faith in a particular technological future provided a valuable and hard-won space in which other scientists and engineers could mobilize, explore, and push the limits of the possible.

The engineering and technical knowledge that underlies visioneers' work is equally critical. O'Neill and Drexler both drew on their academic backgrounds and experience in science and engineering. Their detailed engineering and design studies underpinned speculations as to what the future could be like. Not content with just speculation, O'Neill, Drexler, and others who shared their visions did research to help advance the technologies central to building their imagined futures. Visioneering connects this emphasis on design, engineering, and construction to a more distant time horizon and an expansive view of a future determined by technology.

Even if they didn't have the physical or financial means to build a colony in space or self-replicating nano-assemblers, O'Neill and Drexler depicted evocative worlds through their books and articles. These writings attracted like-minded enthusiasts eager to imagine and perhaps live in these technological futures. Over time, a canon of visioneers' writings developed. These texts helped educate and define visioneers' communities while their supporters used them to launch further debate about what the future could be.

Sometimes, however, the popularity of visioneers' ideas and their writings proved problematic. Space colonies and nanoscale machines became indelibly associated in the public's mind with O'Neill and Drexler respectively. Publishing outside the specialized confines of peer-reviewed scientific journals produced tensions between the necessity of promotion and the inability to control its outcomes. Visioneers' successful popularizations attracted others eager to adopt or co-opt their ideas. Timothy Leary, for example, connected O'Neill-style space settlements to his own vision for how humans might purposefully evolve as a species, much to the physicist's unease.

As journalists paraphrased and repeated visioneers' ideas, some distortion was inevitable. Eric Drexler coined the term "gray goo," for example, to describe a hypothetical scenario in which self-replicating nanoscale machines consume the planet while making copies of themselves, but he gave the idea only glancing mention in *Engines of Creation*. Nonetheless, descriptions of uncontrolled, self-replicating "nanobots"—a term Drexler himself avoided—proliferated. These accounts helped define, for better or worse, how the public imagined a future in which Drexler's nanotechnology existed. Unable to maintain full control of their ideas, visioneers such as Drexler and O'Neill risked ostracism from mainstream researchers who marginalized them to the "freak show that is the boundless-optimism school of technological forecasting."[20]

Like traditional engineering, visioneering requires money. Promoters of radical new technologies create a bricolage of patrons and supporters to finance their work. Many visioneers operated to some degree outside the patronage system that funds and supports conventional American scientists and engineers at universities, corporate labs, or federal facilities. Sometimes this was a deliberate choice, as it provided freedom without managerial oversight and peer review. In other cases, especially with Drexler, the legitimacy that might have come with federal monies and tenured professorships often proved elusive. Funding to pursue and explore O'Neill's space settlements and Drexler's nanotechnology was often ad hoc. Venture capitalists, wealthy entrepreneurs, and curious citizens all contributed in varying degrees. To help raise funds, both men set up nonprofit institutes that helped promote their visioneering. And, although it was less common, federal agencies like NASA and industrial research labs helped support some of the visioneering described in this book. But what primarily motivated both people and organizations to open up their minds and wallets was curiosity and, for some, the chance to get in on the ground floor of something potentially profitable. Like their patrons, visioneers stood to benefit from breakthroughs that helped validate their ideas, with potential rewards coming not just financially but also in the coin of enhanced credibility.

Technology, of course, involves much more than the tangible stuff—the cell phones, freeways, and antibiotics—that shapes our lives and the natural environment. In this book, "technology" represents the diverse ensemble of enabling activities and knowledge as well as the actual "things" themselves. This marks visioneering as something far from a solitary activity. Visioneers engaged in promotion, popularizing, and fund-raising that created and connected different communities while helping advance their broader visions. This heterogeneous engineering created durable social networks while popular explications of their ideas generated wider public interest.[21]

As we survey the overlapping histories of O'Neill's space colonies and Drexler's nanotechnology, we find a few people who proved especially effective at helping get the message out. Stewart Brand, for example, used his magazine *CoEvolution Quarterly* to foster a debate about O'Neill's vision for the future. A decade later, Brand helped raise awareness of Drexler's ideas for molecular-scale manufacturing among business executives and technology pundits. Although not a visioneer as I've defined the term, Brand (and others like him) helped shape public awareness of visioneers' activities and ideas. Over time, the tools for promotion and advocacy changed. In the 1970s, O'Neill's supporters promoted the "humanization of space" with mimeographed newsletters, bumper stickers, and leaflets passed out at science fiction conventions. By the late 1990s, the communities that visioneers fostered could also interact with one another through e-mail, Web sites, and Internet newsgroups. These new tools amplified visioneers' messages as well as the chance of being distorted or attacked.

To sum: *visioneering* means developing a broad and comprehensive vision for how the future might be radically changed by technology, doing research and engineering to advance this vision, and promoting one's ideas to the public and policy makers in the hopes of generating attention and perhaps even realization. Throughout all these diverse activities, people like Drexler and O'Neill worked to build technical and social foundations for their own particular conceptions of the technological future. This book explores what such visioneering entailed, the ways in which it worked, and the places where it went astray.

The histories of science and technology offer other, earlier examples of people we might categorize as visioneers. For example, before World War Two, amateur rocket societies in Germany, the Soviet Union, and the United States blended rudimentary rocket engineering with romantic ideals of a future in which space travel would be routine.[22] Grassroots groups such as the American Rocket Society and the Verien für Raumschiffahrt (German Rocket Society) were inspired by theorists and dreamers, while the drive for space exploration and colonization drew in and drew upon the work of science fiction writers. A considerable amount of charlatanism surrounded these fledgling efforts as well. From this milieu emerged charismatic personalities—visioneers—like Wernher von Braun, Robert Goddard, and Frank Malina, who combined their fascination with exploring new worlds with cutting metal, mixing chemicals, and drawing blueprints. Successfully launching their first rockets also meant selling the dream of spaceflight to the public and potential patrons. Von Braun long imagined ring-shaped space stations and missions to the moon and Mars. His skills at public relations *and* engineering made some of it come true, although at terrible cost to the enslaved workers who helped build Nazi V-2 rockets and the victims of missile attacks in Great Britain and Belgium.[23]

The visioneers' hybrid nature—a combination of futurist, researcher, and promoter—and the influence they sometimes attain compels us to consider how they interact with other actors in broader systems of technological innovation. Business executives and academics have often employed ecological metaphors to describe places where technological innovation occurs.[24] These complex and dynamic "ecosystems" are home to some familiar "species." These include established companies, universities, law firms, patent lawyers, entrepreneurs, investors, government funding agencies, the media, and, of course, scientists and engineers.

But we must also be curious about the visioneers who reside, as did O'Neill or Drexler, in the interstitial niches and edges of such technological ecosystems. Although this book makes no claim to analyze public or industrial policy, it does argue that visioneers, a species typically less acknowledged, are also important to the growth, diversification, and health of today's technological ecosystems. Such an evaluation demands an understanding of the past along with a measured consideration of the role of visioneers in fostering innovation and new ideas. This story also helps explain why utopian-tinted visions of the technological future flourish despite their predilection to mutate, get co-opted for other purposes, or simply disappoint.

Histories of the Future

The scenarios of the technological future this book explores originated with a fascination, even obsession, with the future that surfaced in the 1960s and continued into the following decade. When Senator Edward Kennedy said that "we must be pioneers in time, rather than space," people responded.[25] In 1966, a few people who saw the future as a new frontier formed the World Future Society. By 1974, its membership had climbed past fifteen thousand.[26]

Space exploration, the advent of microelectronics, the growing ubiquity of computers, and the ability to genetically engineer new organisms certainly sparked profound questions about the future. Darker currents in American society also contributed to Americans' tendency to look forward uneasily. Crises of confidence about the government and national power caused by Vietnam and Watergate coupled with inflation, oil shortages, and unemployment made people fearful for the future. One response to this apprehension and anxiety was to try to predict, with an aim toward managing, what the future had in store.

People have always looked *to* the future. But, in the late 1960s, a growing number of scientists, writers, and other experts were also looking *at* the future. Because America's economy, society, and politics appeared unstable, the technological future seemed an especially robust, even hopeful, place for speculation. American businesses started retaining specialists, including science fiction writers, to "plot the future much as medieval monarchs used to have court-astrologers around."[27] Techniques originally developed for Cold War military planning made their way to the corporate world. The growing availability of computers and a belief that complex economic and social situations could be modeled aided their acceptance.[28] The *Limits to Growth* report appeared at the end of a long process, mediated by data and computers, that aimed to discern what the future might hold.

Well into the 1970s, the "future" remained an object of serious scholarly inquiry. Interdisciplinary groups of economists, computer scientists, and sociologists attempted to understand it more "scientifically" and proposed ways for society to navigate toward more desirable futures.[29] Adding to this was the growing community of professional futurists. Paid for their informed predictions, these people gave particular attention to what the key technologies of the future would be.[30] In this golden age of contemplating technological tomorrows, professional "futurologists" became jet-setting celebrities handsomely compensated for their advice.[31] Millions of people bought Alvin Toffler's *Future Shock* (1970) and *The Third Wave* (1980). Toffler, a former editor for *Fortune*, described how modern society was poised between two technological eras, one of industry and another of information, and predicted abrupt changes catalyzed by technologies people needed to prepare for.[32] Futurists like Toffler and John Naisbitt, who wrote the best-seller *Megatrends*, don't qualify as visioneers as I'm using the term. Little detailed technical knowledge underpinned their ideas for the future, nor did they undertake design efforts or engineering to advance the technologies they spoke of.

However, the future that interested the people who populate this book was not next year's business cycle, nor was it some faraway world of the twenty-sixth century. Visioneers' imagined futures, shaped by technologies they helped promote, were just a few decades over the horizon, a time they hoped to personally experience. And even if space colonies or Drexler's nanomachines seemed to be far from realization, they served as political statements. These visions said something as to who was going to build the future, control it, and benefit from it. The future offered a blank space on time's map, a temporal vacuum in which to project one's hopes and fears. Creating visions of the future and the technologies that might help shape it is a political act as well as an exercise of imagination. But the future is not a neutral space. Inevitable disagreements as to what the future will be like and how it might be realized make the future a contested arena where diverse interests meet, debate, argue, and compromise.[33]

We see this dynamic at work today in debates over how to address broad challenges such as climate change. Some people maintain the viability of coal-fired power plants and others advocate renewable energies, while a smaller population envisions radical technologies like fuel cells or nuclear fusion as the power source of the future.[34] Each of these communities envisions, promotes, and works to advance a particular future and the technologies central to it. Influence among those with differing views of the future is rarely symmetrical. Not all futures are created equal. But understanding the history and processes of visioneering can help us get a sense of how we'll write tomorrow's future today.

In this book, we see how different individuals and the groups that coalesced around them vied to construct and claim the future through their writings, their designs, and their interactions with broader publics. In doing so, they often rejected other possible futures, especially those suggesting that the resources of the planet, the ingenuity of its people, and even our own life spans presented limits. To be sure, they also sometimes disagreed with members of their own communities.

Visioneers and the communities of researchers, futurists, and entrepreneurs they attracted often existed at the blurry border between scientific fact, technological possibility, and optimistic speculation. The difference between an eccentric inventor and a visionary entrepreneur is a fine one and often not distinguishable or appreciated at the time. One way to distinguish visioneers' ideas is that while they may have seemed fantastical, they were not impossible. Unlike time travel, designing a space colony violated no obvious physical laws. As one technology enthusiast put it, "Our ideas don't require any new physics to work ... it's just that we follow chains of thinking much further along than most people are prepared to go."[35] As uncomfortable as visioneering may have made some mainstream scientists, no one conclusively exposed its unfeasibility.

As a historian, I am fascinated by past visions of the technological future. I'm less interested in adjudicating whether visioneers' plans for the future were correct or wrongheaded. What concerns me more is how visioneers conceived of and presented their ideas in response to the dire warnings in reports like *Limits to Growth*. What were their motives, hopes, and results? How did other technological communities react to their plans? How were these ideas brought to the public by journalists, science fiction writers, and popular culture? Can we detect ripples from their schemes in the broader American imagination? Although not always visible from the desks of federal science policy makers, I believe visioneers' influence was experienced as frissons of excitement, curiosity, and alarm among other scientists and the broader public. At the same time, my interest is a critical one. As the book's title suggests, visioneers imagined a limitless future, an insistence that can appear naïve. Strictly speaking, there are limits. But the search for how to push past them has been a powerful motive force in the history of technology.

The technologies proposed by visioneers like O'Neill and Drexler existed at the margins of possibility. That was a key part of their attraction. Another clue to their seductiveness stemmed from their potential to solve seemingly intractable social problems. This book places visioneers in the broader tapestry of technological enthusiasm and optimism that has marked much of America's history. Throughout the history of the United States, its citizens displayed a particular flair and fondness for technological utopianism.[36] Part of this stems from the American experience. Early settlers from Europe saw the continent as an unsettled area containing seemingly limitless resources that they could extract and exploit with new technologies. It's no surprise then that twentieth-century visioneers imagined their technologies unfolding in new frontiers—orbiting in outer space, at the nanoscale, off in cyberspace, and so on.

Like the utopian crusaders of the late nineteenth century, modern visioneers wanted to improve society. These modern utopians predicted that technologies they advocated would have a transformative effect on society as humans mastered the ability to create new worlds, undertook atomic-scale engineering, and, if truly successful, overcame their own biological limits. Nonetheless, visioneers and their supporters were not immune to the lures of profit, celebrity, and sensationalism.

The futures depicted by the people in my story did not unfold as they predicted or hoped. Regardless of their reception or indeed their success as judged by today's circumstances, it is important to closely examine the history of these "failed futures." Through them, we see the challenges visioneers faced in conceiving their ideas, trying to implement them, and defending them against critics and rivals. Along the way, they came to terms with what were sometimes partial or even Pyrrhic victories.

Failure, of course, is a subjective term, and visions of the future do not die easily. Moreover, if we study only "successful" technologies, our overall understanding of technology and its history becomes dangerously skewed. Any history of the American automobile that takes into account only the internal combustion engine hides the many other ways cars were powered, not to mention "failed" transportation systems such as trolleys, high-speed rail, or even the much-lampooned flying car. Uncritical acceptance of such a narrative makes cars powered by an internal combustion engines look "natural" when their "success" really was anything but.

One must view the activities of the technology enthusiasts described here in the context of their time, not by the extent of their success so far. These visions of the future were taken seriously by many people *at the time*. A richer historical picture emerges with a more symmetrical appraisal of success and failure. How did the technically possible become a path taken or not taken? Are visioneers' ideas still shaping our conceptions of the future? In this book, we find visioneers as part of a longer chain of technological enthusiasm and optimism that has marked so much of America's history.

This book's main narrative starts around 1972. It is a story that I am writing some forty years later. Yet, in many key ways, 1972 and 2012 seem closer than the passage of four decades might suggest. Today's planetary threats from climate change and overpopulation resemble in many ways existential dangers that *The Limits to Growth* underscored in 1972. Now, as then, economic uncertainties abound as do questions about the abilities and limits of government to address them. And, in the wings, advocates of geoengineering, synthetic biology, fusion energy, cloud computing, and other "emerging technologies" circulate and visioneer their designs for the future. Exploring the activities and ambitions of people such as O'Neill and Drexler adds depth, richness, and nuance to our understanding of today's technological society, how we arrived here, and where we might be going.

CHAPTER 1

Utopia or Oblivion for Spaceship Earth?

This is no way to run a spaceship.
—Kenneth E. Boulding, 1965

When the Club of Rome released *The Limits to Growth* in 1972, it came as the culmination of growing ambivalence, confusion, and pessimism about the future and technology's place in it. In the United States, Americans' reluctant recognition that the planet had finite resources, coupled with fears of uncontrolled consumption and population growth, profoundly shaped this pessimistic context. The roots of this anxiety went deep, though, and tapped reservoirs of anger as well.

In 1964, Mario Savio, a leader in the Free Speech Movement at the University of California, Berkeley, exhorted students to throw their bodies onto the "Machine" as protest. Savio's choice of metaphor was no accident. The figurative gears, wheels, and levers depicted by Savio represented what he and other critics saw as the dehumanizing effects of technology on people and nature. That same year, when President Lyndon B. Johnson visited the New York World's Fair, he extolled the remarkable accomplishments of scientists and engineers. Few people could have imagined a world, Johnson proclaimed over the voices of civil rights protesters, knit together by satellite communications, powered by nuclear reactors, or made healthier through wonder drugs, organ transplants, and agricultural bounty. But, as sodden flags moved limply in the chill April breeze, Johnson intoned that Americans and the planet itself stood at a crossroads of "abundance or annihilation, development or desolation."[1] Despite its enthusiastic, even Panglossian, expressions of faith in the technological future, the Fair's ideological facade showed cracks. Even its motto—"Peace through

Understanding"—struck many as questionable when the global nuclear stockpile topped thirty-six thousand weapons and chemical defoliants, napalm, and push-button warfare were part of an expanding war in southeast Asia. By the decade's end, this general sense of uncertainty was transmuting into widespread pessimism.[2]

Americans' conversion from unabashed enthusiasm for technology to ambivalence and hostility did not occur suddenly, nor can we trace it to a single cause. Some Americans, for example, were concerned about the mortal dangers of the escalating arms race, while others worried about the pollution of the country's skies and waterways or questioned societal values that prized conformity, consumerism, and planned obsolescence. For others, the "plastic fantastic" futures depicted in corporate advertising weren't just banal and boring but appeared threatening. Whatever the direct cause, Americans' overall attitudes toward science and technology became more complex and questioning.

To be sure, not all Americans mistrusted technology. But more people were thinking, writing, and worrying about the future it might help create. They wanted and expected more from technology. In 1968, counterculture icon Stewart Brand anticipated this ideal on the first page of the *Whole Earth Catalog*: "We are as gods and might as well get good at it." By the time the Nixon presidency ended in scandal and resignation, a new language of limits and scarcity replaced expectations of abundance.[3] To begin to appreciate the unconventional solutions that visioneers offered in response to the challenge of limits, we must first look more closely at such sentiments.

Living on a Lifeboat

A major surge of ideas and issues that fed the mounting pessimistic outlook about technology and the future sprang from Americans' concerns about the environment, both in their own backyards and throughout the world. For those who opposed the war in Vietnam, images of bulldozers, defoliated jungles, and bomb-cratered landscapes made the conflict appear as much a war on nature as it was about the indiscriminate use of high-tech weaponry.[4] The thought that the earth was merely a "tiny raft in the enormous empty night" amplified unease about the planet's finite resources and its surging population.[5]

Atomic Age anxiety helped foster growing environmental awareness among scientists and the public. In 1958, while scientists and politicians debated technical details of a nuclear test ban, the Committee for Nuclear Information, cofounded by biologist Barry Commoner, announced it would start gathering and analyzing thousands of baby teeth. The headline-making project demonstrated that radioactive materials from atmospheric nuclear tests were entering both the global food chain and people's bodies.[6] Oceanographers likewise detected the spread, via global ocean currents, of radioactive waste dumped at sea by nuclear superpowers.[7] The publicity such studies received helped show Americans that distant events and actions could have consequences closer to home.

This point became impossible to ignore after the publication of Rachel Carson's *Silent Spring*. Serialized in the *New Yorker* in June 1962 and then presented as a book a few months later, *Silent Spring* forced Americans to question the beneficence of technological solutions. In Carson's case, the issue was the effect of chemical pesticides on wildlife and public health. This was no abstract subject but one that resonated in Americans' own backyards and playgrounds.[8] It's worth noting that Carson herself wasn't antiscience, as she encouraged biological engineering as an alternative to chemical control of insects. Despite such niceties, songs from this era reflected the mounting sense of emergency, from Tom Lehrer's satirical song "Pollution" ("like lambs to the slaughter, they're drinking the water") to Joni Mitchell's folk hit "Big Yellow Taxi," which mourned a paradise paved over for parking lots. Even the site of the 1964 New York World's Fair, planned as a paean to the technological future, received critics' scrutiny, taking place as it did on a former wetland in Queens that engineers had "improved" to the point where it supported little wildlife at all.

Throughout the 1960s, politicians, writers, and activists blended scientists' notions of delicately balanced ecological systems with imagery derived from the burgeoning American and Soviet space programs. The result was a trope that gained in popularity throughout the heyday of U.S. and Soviet efforts to best one another in space exploration: "Spaceship Earth."

The idea's popularization originated with Kenneth E. Boulding, an economist at the University of Michigan. In May 1965, Boulding gave a talk titled "Earth as a Space Ship" in which he contrasted two economic models and their consequences for the future. For centuries, Western civilization had expanded via what Boulding called a "cowboy economy." Unheeding of limits and marked by a focus on consumption, it was prone to "reckless, exploitative, romantic, and violent behavior." Now, with environmental doom just over the horizon, society needed to recognize that earth was a "single spaceship, without unlimited reservoirs of anything." Instead of wasteful consumption coupled to relentless pressure to increase production, this "spaceman economy" must be commensurate with life aboard "a tiny sphere, closed, limited, crowded."[9]

The burgeoning environmental movement adopted the metaphor of "Spaceship Earth," and the term entered the popular lexicon of the 1960s. Adlai Stevenson, for instance, told the United Nations that we are all "passengers on a little spaceship," while the phrase "Spaceship Earth" became the title of a popular book by British economist Barbara Ward. Futurist-designer Buckminster Fuller claimed some ownership of the phrase as well, and in his book *Operating Manual for Spaceship Earth* he urged engineers, scientists, and world leaders to steer the planet away from imminent ecocatastrophe.[10]

To be clear, Boulding didn't insist that the earth actually *was* a closed system. He acknowledged, for example, that sunlight could be converted into solar energy. Nor did Boulding (or some other ecology-minded people) support actual human space exploration. Of the Apollo program's profligate use of the planet's resources, he said in his 1965 talk: "This is no way to run a spaceship." Instead, Boulding and other intellectuals who adopted his concept wanted to emphasize that people needed to think of the planet as *analogous* to a closed system. Adopting such a view would help foster more sustainable economic and social behavior. However, Boulding conceded that establishing this homeostatic "spaceman economy" would require sacrifices. This might mean restrictions on individual freedoms, limiting national sovereignty, or opting for carefully planned economies that placed less of a premium on free enterprise. A decade later, such sentiments prompted forceful reactions from free-market economists, libertarian-minded citizens, and, of course, this book's visioneers.

The idea that the earth might be likened to a spaceship became much less abstract when Apollo astronauts sent back the first color pictures of the whole planet from space. On Christmas Eve in 1968, *Apollo 8* astronaut William Anders captured the planet—white swirls of clouds, brilliant blue oceans, and russet-colored landmasses—emerging from behind the lunar wasteland, a jewel-like disk suspended against the blackness of space. "Earthrise," as the global media christened Anders's photo, became one of the most influential photographs ever composed and an enduring legacy of the Apollo program.[11]

Coming at the end of a year that had seen the 1968 Tet offensive in Vietnam, campus unrest, and the assassinations of King and Kennedy, "Earthrise" provided people with a new perspective of their place in the universe and briefly inspired hopeful sentiments for the future. For writer Archibald MacLeish, the image showed "the earth as it truly is, small and blue and beautiful in that eternal silence," and he hoped it would encourage people to "see ourselves as riders on the earth together."[12] Photos of the earth from space became ubiquitous symbols for the environmental movement; Stewart Brand, who had lobbied NASA to release such photos, put "Earthrise" on the cover of the *Whole Earth Catalog*, his wildly popular publishing experiment.

However, bodings of an apocalyptic environmental crisis still surged underneath this sense of wonder and awe. For many ecologists and economists, the global environmental problem stemmed from a single intractable and pervasive cause: "dedication to infinite growth on a finite planet."[13]

Fears of unchecked population growth already had a long history. In the late eighteenth century, for instance, Thomas Malthus presented a gloomy vision of humanity's future in his *Essay on the*

Principle of Population. Demographic data convinced the English clergyman and mathematician that "premature death must in some shape or other visit the human race" unless restraints on population growth emerged. Read abundantly and attacked widely, Malthus's ideas provided a touchstone for future debates over resources and population and triggered spasms of debate throughout the twentieth century.[14] After World War Two, books like Fairfield Osborn's *Our Plundered Planet* and William Vogt's *Road to Survival* stirred a new "wave of postwar pessimism" by proclaiming that "the Day of Judgment is at hand" because of overpopulation and competition for resources.[15] Consequently, mounting concern about population growth stimulated the creation of organizations like the International Planned Parenthood Foundation and Zero Population Growth. The idea that famines in India and elsewhere might breed social and political unrest suggested a new national security threat, one very different from that posed by Soviet military might. As the United States committed more troops to fight in Southeast Asia, journalists bluntly stated that Western nations would have to decide whether to "feed 'em or fight 'em."[16] Intimately connected with '60s-era fears of uncontrolled population growth were alarmist predictions of dwindling natural resources, which hungry populations and newly developing nations released from colonial confines only exacerbated.

Some academics expressed similarly harsh views. Garrett Hardin, an ecologist at the University of California, perceived direct links between overpopulation and ecological and social collapse.[17] His 1968 essay "Tragedy of the Commons," one of the twentieth century's most widely circulated articles, depicted people's tendency to act out of personal self-interest and, in the process, destroy communal resources and swamp humanity's metaphorical lifeboat. In his grim analysis, Hardin called for more regulation, curtailed immigration, and reduced reproductive choice, extreme prescriptions that, not surprisingly, colleagues critiqued as faintly disguised "barbarism."[18]

Another dramatic warning of overpopulation's perils came from Stanford biologist Paul R. Ehrlich. The opening of his 1968 book *The Population Bomb* places Ehrlich in a crowded Delhi street where the hellish "dust, noise, heat, and cooking fires" of the poor and underfed overwhelm him.[19] Americans, he wrote, must recognize and empathize with their "less fortunate fellows on Spaceship Earth." Otherwise, Ehrlich predicted, millions of people in deprived nations would soon be starving worldwide, and the resulting humanitarian crises would threaten America's own domestic security. While some experts saw such claims as wrongheaded—ecologist Barry Commoner insisted Ehrlich simplified the cause of a much more complex set of social problems—his jeremiad, endorsed by the Sierra Club and promoted by appearances on television talk shows, was read by millions.

Even though scientists had made remarkable progress in increasing food output for the world's developing countries, solutions that relied only on technology offered no absolute answer. To Ehrlich and his supporters, even the vastness of outer space promised no safety valve. "We can't ship our surplus to the stars," he insisted. Some researchers agreed. In 1970, the Norwegian Nobel Committee presented its annual peace prize to Norman Borlaug for his work in helping foster the Green Revolution that was increasing food production worldwide. When the American agronomist accepted his award in Oslo, he reminded his audience that growing more food for a hungry planet was only a temporary solution unless societies curbed population growth and the overconsumption of resources.[20] Failure to do this, warned Ehrlich, Hardin, and scores of other writers, would have catastrophic consequences for the future. But "where," asked Borlaug in 1970, "were the leaders who have the necessary scientific competence, the vision, the common sense, the social consciousness, the qualities of leadership, and the persistent determination" to make such hard choices? A few years later, a small group of influential global citizens stepped forward to offer their own answers.

Data In, Doomsday Out

The Limits to Growth originated in the research of Jay W. Forrester, an MIT professor once described by the *Wall Street Journal* as someone who "likes to play disturbing games with graphs."[21]

Born in rural Nebraska, Forrester moved to MIT in 1940 with the intent of studying electrical engineering. During World War Two, he worked in MIT's Servomechanism Laboratory, where engineers designed complex cybernetic systems to control the movement of gun turrets and radar antennae. After the war ended, Forrester helped design the first high-speed digital computers, getting rich from his innovations in the process.

In 1956, Forrester switched careers and moved to MIT's Sloan School of Management. He intended to apply what he knew about engineering systems, computers, and decision-making processes to corporate management. At Sloan, Forrester developed tools and techniques for modeling and manipulating economies and cities. One of his first projects was studying real-world factories as systems in which variables such as production, inventory, and employment were treated as interconnected streams of information. In the process, Forrester developed a methodology he called "system dynamics." His models typically portrayed what he termed "overshoot and collapse." In such scenarios, outcomes like productivity would rise initially, generating overproduction, and then fall back rapidly, creating factory closures and unemployment. Therefore, a desirable goal was achieving some state of balance and equilibrium. Forrester hoped his mathematical models would become the basis for rational decision making by business owners and policy makers.

In the late 1960s, Forrester broadened his focus from discrete industrial systems to entire cities, hoping to rationally tackle the processes associated with the decline of urban areas. Some of his computer-derived models for "urban dynamics" suggested, for instance, that low-cost housing actually increased poverty by encouraging poor and underemployed people to congregate in cities, which then suffered lower tax revenues. Some reviewers saw such conclusions as insensitive in the politically charged environment of the late 1960s, while others critiqued Forrester's tendency to eschew reams of empirical data or the work of urban planning experts in favor of computer-based models.[22]

Similarities existed between Forrester's goals of making models to guide policy and how researchers from other fields had begun to view the world. Before World War Two, economists like John Maynard Keynes used the flow of money and the response of producers and consumers to create large-scale models of how national economies function. The inclination to think in terms of large dynamic systems was also adopted by biologists who proposed that ecological systems could be understood as interactive entities governed by feedback loops. Just as macroeconomists based their studies on the flow of money, systems ecologists of the 1950s and 1960s assumed energy, as converted by organisms, was the basic resource in circulation and applied cybernetic principles to describe the workings of natural features such as watersheds and ocean atolls.[23]

Throughout the 1960s, researchers developed an array of computer models to describe the behavior of subatomic particles, industrial production, traffic flow, defense networks, and climate patterns.[24] The next step seemed obvious—could one model the entire world, and if so, what might this reveal?

The tools developed by Forrester and his colleagues promised to reveal the inner workings of global social and economic systems. The Club of Rome decided that these techniques might help it better understand what it labeled "*le world problematique*." By this, the Club of Rome's members meant the ensemble of "tangled, changing, and difficult problems" that nations around the globe faced in varying degrees.[25]

The Club of Rome got its start in 1967 as the result of a meeting between Alexander King, a British chemist and director general for the Organization for Economic Cooperation and Development, and Aurelio Peccei, a well-connected Italian businessman interested in global development. Their encounter was brokered by Carroll L. Wilson, one of Forrester's colleagues at MIT and "an activist on the world stage."[26] King and Peccei envisioned creating a group similar to the Lunar Society, the famed group of entrepreneurs and intellectuals that had met regularly to promote science and industry in eighteenth-century Britain.[27]

King and Peccei decided to bring friends and colleagues together so they could discuss potential solutions for the world's major problems. But, rather than trying to address environmental, poverty, and political issues in isolation, the Club of Rome, as they christened themselves, wanted to treat these challenges in a systemic and scientific way. However, getting a meaningful perspective on such a wide range of issues proved difficult. Corporate patrons rebuffed the club's initial appeals because it lacked a suitably rigorous approach. Having "no methodology and no money," the club turned to Jay Forrester and MIT.[28]

In June 1970, Forrester met with the Club of Rome in Bern, Switzerland. He suggested that his "systems dynamics" approach might help them tackle the complexity of the *problematique* and see connections between such seemingly isolated topics as commodity markets, ecological systems, and urban crowding. Forrester proposed building a comprehensive model that cut across academic disciplines to include economic, cultural, technical, and political dimensions.[29] When he returned to MIT, Forrester and his team worked at a frantic pace to create an initial computer-based representation.

Two of the people who helped develop the simulation were Donella Meadows, a Harvard-trained biophysicist, and her husband Dennis, a young researcher at MIT who had received his degree under Forrester's guidance. The couple started by modeling global interactions between major subsystems like agricultural production and population growth. Three weeks later, Forrester's team unveiled its prototype model, which it named WORLD 1, to club members.

Peccei and Club of Rome members examined its outcomes with horrified fascination. Despite sketchy data inputs and modifications to variables like population, food production, and industrial output, each iteration of the computer model gave the same sobering prediction: "disaster at some time in the not-too-distant future."[30]

Now possessing a viable set of tools in the form of Forrester and Meadows's computer- subroutines and models, the club successfully garnered $250,000 from the Volkswagen Foundation to probe the issue more closely. It also asked the MIT team, supervised by Forrester, to build an even more robust version of the original WORLD model. Whereas Buckminster Fuller, Kenneth Boulding, and others had used Spaceship Earth metaphorically, the

MIT researchers were trying to mathematically conceptualize the planet with its ebbs and flows of various of resources and refuse. In a sense, their work aimed to provide a quantification of the "global picture" captured in NASA's famous "Earthrise" photo.

Using feedback principles, automatic data processing, and lots of mainframe computer time, the MIT team constructed and ran their global model, the first with a "time horizon longer than thirty years," to investigate five key interconnected trends: accelerating industrialization, population growth, increasing malnutrition, resource depletion, and environmental loss. The goal was not to predict the future per se but to generate a range of simulations to help see how "different modes of behavior over time" might generate changes, positive or disastrous, throughout the global system."[31] At its heart, *Limits* was most concerned with the dangers of uncontrolled, exponential growth in deleterious variables such as population and resource use.

As Forrester had done with his earlier systems modeling, Dennis and Donella Meadows opted to focus on exploring the relations *between* variables instead of the quality of the data itself. The complex model called WORLD 3 that the group produced was admittedly "imperfect, oversimplified, and unfinished."[32] For example, in 1970, with the World Bank and the United Nations only a quarter century old, robust data over the timescale that the MIT group needed didn't exist or wasn't easily collated. Another problem, at least for potential critics, had to do with the group's a priori supposition that, as their proposal to Volkswagen said, "growth cannot continue indefinitely on a finite planet . . . we are faced with an inevitable transition from world-wide growth to global ecological equilibrium."[33] WORLD 3, which became the basis for the predictions in *Limits*, assumed that certain positive variables, such improvements in technologies, would grow in fixed linear fashion. However, they assumed other variables, such as population, pollution, resource use, would expand exponentially. To detractors, this sounded like the MIT team began its study with its basic conclusions predetermined.[34]

In April 1971, at an opulent estate in Ottawa, Dennis and Donella Meadows presented preliminary findings from their new

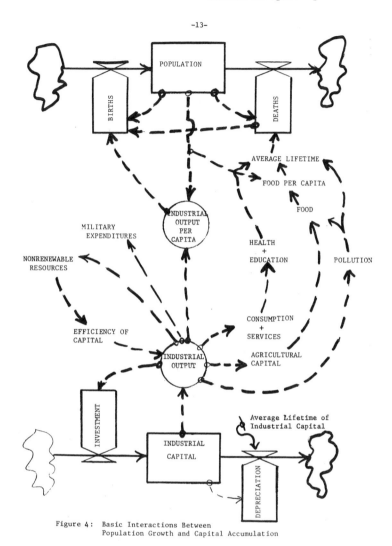

-13-

Figure 4: Basic Interactions Between
Population Growth and Capital Accumulation

Figure 1.1 Flowchart from the MIT–Club of Rome "Project on the Predicament of Mankind"
prepared by Dennis L. Meadows, January 1972. This figure, showing the interactions between
population growth and capital accumulation, was typical of the relationships the *Limits to Growth*
report hoped to reveal. (Image from Jay Wright Forrester papers, MC 439, box 54, Massachusetts Institute of Technology, Institute Archives and Special Collections, Cambridge, MA.)

model to the club's members. The response was underwhelming. "They didn't get it," Donella Meadows recalled as the MIT team had to explain concepts like exponential growth or the finite limits of resources to some participants.[35] Wanting to get their message across more effectively, she produced a bowdlerized version that eliminated references to cybernetic feedback loops or complex computer printouts. Instead, she and Dennis emphasized their fundamental point—their computer-derived model concluded that continued exponential growth would result in a global collapse sometime in the twenty-first century.

By this point, however, Aurelio Peccei had seen enough to be convinced. Over the next six months, he and other club members orchestrated a global publicity campaign in advance of releasing *The Limits to Growth*. Even before it appeared, there was strong media interest as reporters clamored for information about the Club of Rome and its "computer curve to doomsday."[36] Curiosity about *Limits* was heightened further when the *Ecologist*, a British magazine, published "Blueprint for Survival." Like *Limits*, this report called for an economically and ecologically stable society and garnered headlines throughout Europe and the United Kingdom.[37]

Ironically, to produce its no-growth, ecocatastrophist message, the group at MIT had used tools and techniques derived from years of work Forrester had done on behalf of military and corporate clients. The timing for *Limits* was right, though. The MIT group did its predictive modeling in an atmosphere rife with apprehension about technology and the environment and at a time when there was an upsurge in "future studies" by celebrity futurists like Alvin Toffler and Herman Kahn. Concern and interest about the future dovetailed with fin de siècle musings as the twentieth century drew to a close.

Fears about the future were not just the province of activists, campus intellectuals, and ecologists. In the early 1970s, the Christian fundamentalist revival in the United States coincided with apocalyptic excitation about the future. In *The Late Great Planet Earth*, former tugboat captain Hal Lindsey used his own interpretation of the Book of Revelation to write what became the best-selling nonfiction book of the 1970s. Even here, technology was implicated as Lindsey prophesied that new tools like the "world-wide computer banking system" and "in-home computers" would allow the Antichrist to control the future.[38] With an eye toward profit, other evangelical and survivalist writers penned similar visions of the doomed future.[39] When Lindsey's book was made into a movie narrated by Orson Welles, who first found widespread fame with his apocalyptic *War of the Worlds* radio broadcast, it featured people like Ehrlich and Peccei who agreed, albeit for different reasons, that humanity stood on a precipice. While Forrester, Meadows, and the Club of Rome were of a different stripe than evangelical doomsayers, the era's confluence of technological pessimism, environmental concerns, and trepidation about the future shaped both how *Limits* was written and the ways in which people reacted to it.

Debating the Scarcity Society

Although rumors about *Limits* were already circulating among policy makers and reporters, the Club of Rome's report, released as a small book, wasn't formally unveiled to the public until March 2, 1972. Hyperbole also preceded the publication of *Limits*.[40] For example, in invitations preceding the book's release, James Killian, a former MIT president and Eisenhower-era science adviser, was quoted as saying the eighteen-month study was the "most important work" his school had produced in half a century.[41] As a book, *Limits* was largely written by Donella Meadows and published by a hitherto unknown Washington think tank called Potomac Associates. To mark its release, the Woodrow Wilson International Center for Scholars held a symposium in collaboration with the Club of Rome and the Smithsonian Institution. The Xerox Corporation underwrote the gala affair, and a Washington public relations firm was hired to promote the "intellectual bombshell." For those elites who couldn't attend, the Club of Rome mailed some twelve thousand copies of *Limits*.[42]

Scores of politicians, business leaders, diplomats, and journalists converged on the National Mall for the *Limits* symposium.

Underneath the towering columns of the Smithsonian's gothic Castle and surrounded by busts of stern-faced nineteenth-century scientists, they listened while Dennis Meadows presented the report's basic findings. Using large charts covered with soaring and plunging colored curves, Meadows warned about the perils of continued and exponential growth. Meadows, then just twenty-nine years old, took pains to note that his group's intent was not to predict the future but rather to simulate and model how various decisions and behaviors would affect the planet in the future.

The authors of *Limits* claimed their work challenged those "people who speak about the future with resounding technological optimism."[43] But Meadows also insisted he and his group were not opposed to technology. "We are technologists ourselves," he said, "working in a technological institution." However, technical solutions "designed to reduce some of the pressures" that economic and population growth caused "can serve only to postpone the collapse." New technology alone wasn't the answer unless it came accompanied by radical changes in social, economic, and political behavior. Otherwise, like Forrester's initial studies, the computer runs underpinning *Limits* predicted catastrophic collapse for one reason or another sometime in the twenty-first century.

The only long-term solution, according to *Limits*, was an equilibrium state in which economic and population growth were actively stabilized. An orderly transition to this new "scarcity society" would demand a firm hand at the tiller of Spaceship Earth. This meant, among other things, extensive management and regulation of politics, societies, and economies.[44] Otherwise, failure to address critical issues such as overpopulation, pollution, and resource shortages might result, as Garrett Hardin had predicted four years earlier, in the "need to reexamine our individual freedoms to see which ones are defensible."[45]

Critiques of the business-as-usual futures that *Limits* foretold began even before janitors had turned off the klieg lights inside the Smithsonian's Great Hall. Dissection of the book continued for months as journalists, economists, and scientists offered their own interpretations in scores of opinion pieces, essays, reviews, and follow-up studies. As might be expected with such a well-publicized and controversial set of conclusions, reactions diverged widely.[46] At one pole, liberal intellectual Anthony Lewis praised *Limits* as "one of the most important documents of our time" and featured it in a sobering series he called "To Grow and Die," which ran in the *New York Times*.[47] Many writers agreed that the book's essential message affirmed the need for more environmental protection, investments in appropriate technologies like wind and solar power, and, as small-is-better advocate E. F. Schumacher phrased it, promoting "economics as if people mattered."[48] There was a sense that perhaps the environmental collapse *Limits* forewarned could "achieve for businessmen what the atomic bomb gave to physicists—a heightened sense of anxiety about the social utility of their work."[49]

Other readers weren't so charitable. *Science* labeled it a "readable little book for laymen that may prove as popular as Linus Pauling's recent treatise on vitamin C," a verdict that accorded *Limits* scant praise. A review in the *New York Times* savaged the book as an "empty and misleading work" that concealed an "intellectual Rube Goldberg device."[50] The *Wall Street Journal*'s editorial pages noted that *Limits* failed to realistically account for basic economics such as the belief that people consume less when prices go up.

Economists who assumed growth was a fundamental tenet of modernity proved particularly hostile to *Limits*. Demand for resources, critics said, was historically contingent. Factory owners did not clamor for coal in the sixteenth century, nor was uranium a desired international commodity until after 1945. Technological innovation, meanwhile, remained unpredictable and surprisingly generous. As a result, *Limits* appeared to some people as "more repetitious than striking," a reprise of old Malthusian ideas.[51] Simply extrapolating existing trends, which critics claimed already rested on a shaky foundation of data, into the future while allowing no role for human intervention or game-changing technological innovation weakened the credibility of *Limits*. Had MIT's study been done with nineteenth-century data, skeptical economists said, when growth of urban populations coincided with increased numbers of horse-drawn vehicles, the computer would have spewed out a prediction of cities buried under horse manure.[52] A smaller group of scholars counseled a more optimistic view of the future, saying "unlike resources found in nature, technology is a manmade resource whose abundance can be continuously increased."[53] Meanwhile, universities and think tanks throughout the United States and Europe generated their own studies to reexamine *Limits*'s empirical claims.[54]

The passionate condemnation from academic experts turned personal and, at times, nasty. "Irresponsible nonsense," claimed an economist, while one of Forrester's colleagues at MIT labeled *Limits* a "messianic" endeavor by a group of people who "want to save the world." *Limits* merely offered "simple-minded answers for simple-minded people who are scared to death."[55] The book's release in the absence of peer review with a public relations extravaganza did not help assuage skeptics. Moreover, the data ("opinion rather than proved fact") underlying *Limits*'s conclusions weren't available for inspection, which aroused scientists' suspicions further.[56] Even the analytical tools the MIT group used provoked ire. The computer cliché "garbage in, garbage out," although still relatively novel in 1972, typified many experts' responses as critiques with titles like "Caveat Computer" or the "Computer that Printed Out W*O*L*F." appeared.[57]

Debate extended beyond issues of methodology to broader implications. One of President Nixon's cabinet secretaries praised the MIT group for their good intentions but wondered how nations could halt growth without inviting "the destruction of our liberties and freedom."[58] If one embraced the implications of the "Spaceship Earth" metaphor, this meant accepting "the strictest sort of economic and technological husbandry." Such choices would make the political future "much less libertarian and much more authoritarian." It meant, in other words, the end of "freedom as an infinite resource."[59] Even as it was debated in the United States, *Limits* suggested very real consequences overseas. For leaders of countries like Mexico and India, which were in the process of technological modernization, a future with a steady-state global economy appeared as "chilling as the doomsday prophecy . . . an elitist aristocratic, white-man's conspiracy" to lock developing "nations into perpetual poverty."[60] Others saw *Limits*'s proscriptions differently. Chinese policy makers, for example, adopted methodologies similar to what the MIT team had used to formulate their country's "one-child" policy.[61]

Despite the fierce debate, almost all participants recognized, as the pro-growth weekly *Business Week* said, "on a finite planet, growth must end sooner or later."[62] The question was *when* (and how and for whom), realities the Club of Rome recognized in more nuanced and less hyped follow-up studies.[63] Nonetheless, ideas from *Limits* percolated into political discourse. When Nixon, for example, addressed an audience of graduating college students, he criticized "Malthusian pessimism about the future" but also grudgingly called for some compromise between "the Cassandras and the Pollyannas."[64]

Limits received even more attention in June 1972 when, after years of planning, the United Nations convened its Conference on the Human Environment in Stockholm. For twelve days, thousands of delegates from some 114 nations and hundreds of nongovernmental groups met for the first time to discuss global environmental issues. *Time*, recognizing the hundreds of youthful idealists in attendance, dubbed the meeting "Woodstockholm." *Limits* shaped the conference's agenda and discussions as press coverage of the event noted that debates among attendees often revolved around the question of whether a "steady-state world" was viable.[65] The seemingly endless debates at Stockholm made it clear that any progress would be slow. "Each delegation," one attendee said, "consists of an environmental minister, and behind him sits a scientist telling him what to say and a diplomat telling him not to say it."[66] Observations such as this undoubtedly helped reinforce the views of people like Gerard O'Neill that solutions to social and environmental problems would be found not at the negotiating table but in new technologies.

Newspapers and magazines revisited the MIT study and its grim conclusions again and again, making the idea of "limits" a leitmotif for the decade. Current events certainly contributed to this. Eighteen months after the public encountered the *Limits* report, the Organization of Petroleum Exporting Countries (OPEC) announced its first oil embargo. The shock coincided with major upheavals in the U.S. stock market. Oil prices shot up, and American bureaucrats enacted new rules to govern a scarcity society. This meant gas rationing, year-round daylight saving time, and national speed limits. America appeared to be, as the cover of *Newsweek* put it, "Running Out of Everything."[67] President Jimmy Carter internalized themes from the Club of Rome's report in his inaugural address, saying that "even our great nation has its recognized limits . . . we can neither answer all questions nor solve all problems." After Carter failed to win a second presidential term in 1980, he recalled that "dealing with limits" had become "the subliminal theme" of his presidency.[68] Fears of future social collapse à la *Limits* and a distaste for modern technology spurred many anxious people to leave vulnerable urban areas for the supposedly greater safety of rural homesteads or encampments. It is no coincidence that Ted Kaczynski (aka the Unabomber) abandoned his Berkeley professorship for a remote cabin in Montana or that survivalist magazines like *Soldier of Fortune* appeared on newsstands around the time that *Limits* was receiving widespread attention and debate.[69]

The basic tenets of *Limits* also turned up repeatedly in pop culture's ephemeral marketplace. A generation that had grown up in the shadow of the Bomb's "unremitting banality and inconceivable terror" and given its baby teeth for studies of radioactive fallout encountered scores of books and articles containing piercing words like "survival," "doomsday," "crisis," and "ecocatastrophe."[70] While activists formed groups like Friends of the Earth and Zero Population Growth to combat present-day ills, writers and movie directors were busy showing disaster-ridden images of a "futureless future."[71] As "alarmism became something like official federal policy," many American citizens saw a decade "bathed in a cold Spenglerian apprehension" with "the future as a possible enemy."[72]

Science fiction, of course, had long grappled with dystopian depictions of the future. During the 1960s and into the 1970s, science fiction books and films explored dystopian worlds of corporate control, restrictions on individual freedoms, and resource shortages with an enduring frequency.[73] Harry Harrison's novel

Make Room! Make Room! reached a much wider audience in 1973 when MGM turned it into the film *Soylent Green*. Although the film's closing scene, with Charlton Heston's anguished cries and clenched fist, has since become campy humor, its depiction of an overcrowded and overheated New York presaged a decade of profitable Hollywood blockbusters that turned technological catastrophe into uneasy entertainment.[74] Writers set many other books, films, and shows—*A Clockwork Orange* (a 1971 film made from a 1962 book), *Doomwatch* (a popular BBC television series in the early '70s), *Silent Running* (1972 novel), *The Sheep Look Up* (1972 film)—in similar imagined futures ravaged by technology and environmental destruction.

Even the youth counterculture, with its "don't trust anyone over thirty" slogan, was targeted. *Logan's Run*, a 1967 science fiction book turned into an Oscar-nominated movie, describes a hedonistic and corrupt youth-oriented society where life spans have strictly imposed limits. With color-changing crystals implanted in their palms, the "survivors of war, overpopulation and pollution" live for pleasure in their giant, domed city, "freed by the servo-mechanisms which provide everything" until "Last Day," when their crystals blink and they must report for a fiery ritual of self-destruction.[75] Some refuse to submit to that dictate of death and flee. With its pervasive and oppressive technosystems and environmentally shattered world, *Logan's Run* encapsulated two decades worth of mounting ambivalence and pessimism about technological modernity and ecocatastrophe.

The theme of limits permeated academic monographs and countless magazine articles as well as science fiction films and songs throughout the 1970s. Meanwhile, images of lifeboats and Spaceship Earth captured the attention of the public and of policy makers in the 1970s. They also helped inspire visioneers and other technology enthusiasts who flocked to them in search of alternative and seemingly radical solutions. Consider the end of *Logan's Run*. The book's hero and his fellow "Runners" finally manage to escape the homicidal "Deep Sleep Operatives" chasing them. Their sanctuary? An abandoned space colony.

CHAPTER 2

The Inspiration of Limits

||||||||||||||||||||||||||||||||||||||||||||||||||||||||||||||||||||||||||||||||||||||||||||||||||||||||||||||||||||||||||||||||||||||||||||

To be pleased with one's limits is a wretched state.
—Johann Wolfgang von Goethe, from *Maxims and Reflections*

In 1969, as autumn approached, Gerard O'Neill was preparing to teach Princeton University's introductory physics class. The school catalog's description—"a course in general physics offering good school preparation . . . two lectures with demonstration experiments . . . Mr. O'Neill"—wasn't terribly inspiring for the scientist, however.[1] As he made his syllabus, O'Neill decided he wanted to take a different approach to the topic. Campus attitudes, he observed, reflected widespread "disenchantment with the sciences" and a "revulsion against authority and against technology." Even his best students seemed defensive and worried about being "accused by their colleagues of being irrelevant" or becoming cogs in corporate or military research programs.[2]

O'Neill had an idea, though. A few weeks before the semester was scheduled to begin, the first American astronauts landed on the moon's surface. "The course had never had a theme as such," he recalled, "so I chose the Apollo project rather than the classical physics problems of pushing frictionless elephants up inclined planes."[3] A topic like space travel offered O'Neill a pedagogical gold mine. A physicist could use it to teach all sorts of basic concepts such as Newton's laws, energy, gravity, and momentum. Moreover, describing the science of spaceflight would allow him to broach more sophisticated ideas like celestial mechanics and perhaps even discuss the physics of managing a lunar landing through basic computer simulations.

Out of the three hundred or so students who enrolled in Physics 103, a few showed an especial proclivity for the topic. For them,

O'Neill organized an extra seminar. He planned to present them with a series of large-scale engineering problems that would encourage them think through the physics involved as well as broader social or economic implications. "I felt that, despite the bad times," O'Neill said, "improvements in the human condition could be reached by using science and engineering in the right ways as opposed to the wrong ways."[4] O'Neill insisted that solutions devised by the students could not rely on technologies "beyond the level of the 1970s or early 1980s."[5] No faster-than-light travel, no magic materials—just the application of knowledge available then or what could be reasonably predicted by modest extrapolations.

O'Neill gave the first topic for discussion to his seminar students in the form of a question: "Is the surface of a planet the right place for an expanding technological civilization?" They never got to the next problem.

After he became famous, journalists often asked O'Neill why he had picked *this* particular exercise to explore with his students. Although he never formulated a single prime cause, he provided enough clues in talks and interviews such that something like an answer can be pieced together.

Exploration, whether to the arctic regions, oceans, or outer space, fascinated O'Neill since his childhood. He maintained a lifelong enthusiasm for what were the cutting-edge technologies of his youth such as radio and airplanes. As an adult, he avidly read and watched science fiction in which space settlements were depicted in exciting yet believable terms. The real-life successes of the Apollo program nurtured O'Neill's interest in space. And, even in his personal life, he remained open to new possibilities. This may have especially been the case when he first started thinking seriously about space settlements. O'Neill—recently divorced and helping care for his three children—met Tasha Steffen in the fall of 1969. Tasha, a striking young woman with long red hair, had just arrived from Germany to work as an au pair for the Gallup family (of Gallup poll fame). A courtship blossomed. They married four years later and eventually had a son.

But, ultimately, these reasons only hint at *why* he thought about space settlements in the first place. Many people had already done

Figure 2.1 Gerard and Tasha O'Neill in 1976, a few years after they married. (Image courtesy of Tasha O'Neill.)

that. What set O'Neill apart was *how* he approached the question of building settlements in space. He treated the entire topic as a rational and detailed engineering problem. But, once satisfied with the plausibility of his concepts, he didn't simply toss his tablets of sums and sketches into a desk drawer and go back to the physics research that he had built his career on. Instead, he steadfastly improved his designs and, just as importantly, promoted his vision for the future to colleagues and members of the public.

Speculator

Gerard Kitchen O'Neill was born in 1927 in New York City. Boats, trains, and especially airplanes occupied a good part of his childhood imagination, as has long been true for many young boys, and transportation technologies would continue to fascinate him throughout his life. His parents, Edward and Dorothy, had met on an ocean liner when his father was returning home from France after being wounded in a gas attack during World War One. O'Neill later boasted that, although just an infant at the time, he had had the chance to see Charles Lindbergh when the aviator made a triumphant return back to New York City.[6]

When Gerard was about seven, Edward O'Neill needed to recuperate from a major surgical operation. The small family—O'Neill had no siblings—moved to a village in upstate New York. Aptly named Speculator, it was nestled deep in the Adirondack Mountains some seventy miles north of Schenectady. With just a few hundred residents, Speculator was a rustic environment for the O'Neills compared with Stamford, Connecticut, where they had lived previously for several years.

With few friends his own age, O'Neill spent a good part of his formative years tramping around Speculator's woods or reading books his parents bought for him. Stories that featured scientists and builders, such as the *Tom Swift* or the *Young Engineers* series, held his attention. O'Neill's interest in engineering, science, and technology persisted in his teenage years. A visit to the 1939–40 New York World's Fair helped forge O'Neill's lifelong conviction that technology was perhaps the most important force to bring about social and economic change. The fair's depictions of a "clean, exciting, fast-moving world, sleek and streamlined" provided the young teen with powerful images of the technological future and left O'Neill with "a sense of the possibility of change inherent in the development of new technology" as well as its "inevitability."[7]

Eventually the elder O'Neill recovered, and the family moved south to a town in the Hudson River Valley. His father, a successful attorney, fought political corruption in Albany and championed urban reform while his mother volunteered for groups like Planned Parenthood. "I was brought up in an atmosphere," their son recalled, "in which it was proper and appropriate to work on large-scale societal problems."[8]

In February 1944, immediately after his seventeenth birthday, Gerard O'Neill volunteered for military service. He hoped to train as a pilot, but minor eye problems torpedoed this plan, so he opted instead to become a radio operator for the navy. Commissioned as a radio technician, O'Neill learned the latest in radar and sonar techniques and, in the summer of 1945, he shipped out to the Pacific theater. The war ended before O'Neill saw any action and, a year later, he returned home and enrolled at Swarthmore College, outside Philadelphia. Motivated by his technical experience in the navy, O'Neill studied physics and math, graduating Phi Beta Kappa in 1950.

At this point, the young graduate had to decide whether to enter the burgeoning technical workforce—with the Cold War well underway, people with O'Neill's skills and knowledge were in high demand—or continue his studies. Opting for the latter, he arrived at Cornell University in the fall of 1950 to study physics, aided by a fellowship from the Atomic Energy Commission.

O'Neill decided to become a physicist at an extraordinary time in the history of American science, circumstances that surely conditioned his sense of what engineering and technology could accomplish. In the 1950s, before pronounced ambivalence about science and technology took root in American society, powerful atom smashers, nuclear-powered submarines, and rockets probing the limits of the earth's atmosphere symbolized technology's transformative potential and the exploration of new frontiers. The prestige and social stature of scientists rose dramatically as well while O'Neill was in college. While he learned the esoterica of high-energy and particle physics, scientists not much older than O'Neill were suddenly visible in circles of power as they gave advice on social policy, arms control, and military strategy. At the same time, the federal government, especially the military services, had a near-religious conversion to the value of scientific research.

In 1954, not quite four years after starting his doctoral program, O'Neill graduated from Cornell. Married and soon a father, he accepted an assistant professor position at Princeton University. His star rose quickly. Energetic, focused, and ambitious, he embarked on a career path that quickly took him into the realm of postwar "big science." High-energy physics was one of science's most competitive fields, and a successful career demanded laboratory acumen balanced with managerial abilities, fund-raising talents, and a willingness to work with large teams of engineers and technicians.

Soon after he arrived at Princeton, colleagues asked O'Neill to help design a new particle accelerator that the Atomic Energy

Commission had recently funded. Previous accelerator designs, like the cyclotrons Ernest Lawrence and his students had perfected at Berkeley in the 1930s, used electromagnetic fields to accelerate protons or electrons to incredibly high speeds before smashing them into a fixed target. Scientists would then sift through the shower of subatomic wreckage looking for new types of particles and teasing out evidence of what the material world was ultimately made of.

In 1956, O'Neill thought of a new way to improve the performance of physicists' particle accelerators.[9] Instead of having a stream of particles collide with a stationary target, O'Neill realized that having two beams themselves collide head-on would be more energetic and efficient. O'Neill suggested that particle accelerators include devices—what later became known as "storage rings"—in which particle beams could be kept circulating long enough for collisions between them to take place.

In mid-1956, however, other scientists were also starting to think along similar lines, so O'Neill rushed to get his ideas out to the physics community.[10] The initial response was lukewarm as many older physicists maintained allegiance to proven designs. Nonetheless, O'Neill insisted that his "numbers came out right" and the device he imagined would work.[11]

The main technical problem O'Neill needed to overcome was steering beams of subatomic particles from the accelerator into the storage rings and then keeping them there without a ruinous loss of their momentum or density. Between 1956 and 1958, O'Neill and a small group of students at Princeton built a laboratory model of a switch that could change the particle beam's path while preserving its quality. This device, called a "delay line inflector," proved O'Neill's engineering ideas sound, and Princeton rewarded him with tenure at the relatively young age of thirty-two.

Late in 1958, O'Neill and his colleagues received the first installment of what would eventually be several million dollars of federal monies to build a full-size version at the Stanford Linear Accelerator Center in California.[12] Each of the two storage rings at the SLAC machine would be about ten feet in diameter. Inside a stainless steel chamber sporting thick glass windows and gold-plated gaskets, beams of electrons would circle through an ultrahigh vacuum millions of times every second. Every so often, one electron would hit another. Detectors located nearby would record the tracks of charged particles created by these collisions, allowing physicists analyze the interactions.

As the project's main designer, O'Neill maintained a brutal schedule of teaching at Princeton, regularly commuting to Stanford to oversee development while trying to sell his storage-ring concept to other labs in the United States and Europe. He later joked that his main research tool during this time was the jet airplane. At times, his bicoastal work situation frustrated O'Neill, who believed that West Coast collaborators took credit for his ideas in his absence while some colleagues at Princeton asked why he simply didn't do research closer to home.[13]

Despite doing years of research on the fundamental building blocks of matter, O'Neill was, at heart, happiest building machines, seeing how they worked, and thinking about how to improve the design. Getting storage rings to work took longer than O'Neill and his coworkers expected. The project, however, taught O'Neill important lessons, from the details of magnet design and high-vacuum techniques to securing grants and the ups and downs of managing a large engineering project. All these skills came in handy when he started to think about the technical and logistical hurdles needed to build space colonies.

Finally, in 1965, O'Neill's machine recorded its first electron-electron collisions. This demonstrated conclusively that storage rings and colliding beams worked as O'Neill had predicted. In the years that followed, scientists at laboratories such as CERN near Geneva incorporated O'Neill's innovation in their own designs for major particle accelerators. In a sense, O'Neill's championing of storage rings proved almost too successful. He was part of a culture that rewarded experimentalists and theoreticians and ruthlessly separated them from instrument makers. Building machines to do physics experiments was not seen as having the same prestige as theory or data-driven physics itself. While some scientists saw instruments as simply a means to do research, O'Neill remained enthused about conceiving and building large-scale projects in their own right. His department chair saw this tendency too, describing him as an "extremely hard worker . . . [who] tends to engage in very large and difficult tasks that require much time to mature."[14] This assessment would, it turned out, accurately describe O'Neill's eventual commitment to promoting human settlements in space.

In the mid-1960s, O'Neill, now in his forties, started to reevaluate what he wanted to do with the rest of his career. While a successful professor in a top-notch physics department, he started to recognize that his interests were no longer in pure research anymore. Moreover, he worried that esoteric studies of subatomic particles were "likely to become even more irrelevant to the rest of the world."[15] "I just don't get as excited about scientific discovery per se," he said after leaving basic physics research, "as I do about the idea of creating something. . . . If I have a talent, it is probably in the area of large-scale systems design."[16] His faculty file reflects this. Princeton's evaluations praised how O'Neill "energetically and selflessly" contributed to design and engineering endeavors, as well as departmental teaching, but noted that he published fewer articles than one might expect.[17]

O'Neill started considering new career options. In 1966, when NASA announced openings for its Scientist-as-Astronaut program, O'Neill was one of some nine hundred applicants. In June 1967, he traveled to Houston for an extensive series of interviews and examinations.[18] But when NASA announced its final selection of eleven people, O'Neill didn't make the cut. Disappointed but undaunted, the physicist resolved to pursue his growing passion for space exploration in another way.

Imagining Habitats for Humanity

O'Neill's proclivity for ambitious engineering projects, his basic physics knowledge, and his curiosity about space exploration all converged in the growing interest he had in space settlements. O'Neill claimed he had long possessed a "desire to be free of boundaries and regimentation." Grand technological projects like the mass migration of people into space increased the chance that people might choose "peace rather than war, diversity rather than repression, human simplicity rather than inhuman mechanization."[19] He found repulsive the descriptions of the near-totalitarian ways that Spaceship Earth would have to be managed in order to stave off disaster. The dismal futures forewarned by *Limits* and other reports of a "steady-state society, ridden with rules and laws" failed to meet expectations O'Neill harbored since a child of a future made exciting and efficient by technological innovation. It also sounded "like a hell of a world" to leave to his children.[20] In fact, the futures described in books like *Limits* and *The Population Bomb* were a *necessary* catalyst for O'Neill's visioneering. Their alarming predictions provided him with a dire view of the future against which he could contrast his own compelling and more desirable vision.

O'Neill saw "three possibilities for a civilization that gets to about our stage," he later told an interviewer: "One is stagnation, one is annihilation, and the third is expansion out into space through space colonies."[21]

Space colonies?

O'Neill was convinced that an ambitious program for humans in space would counter the pessimism, irrationality, and cultural stagnation he feared would continue to grow with society's rejection of science and technology. On the more practical side, O'Neill believed that settlements, solar power facilities, and manufacturing based in space could provide energy and resources for an expanding human civilization while moving off-planet the environmental degradation that industrial activities caused. New challenges, new technological opportunities, and access to boundless resources, to be achieved while shifting humanity's environmental footprint caused by overpopulation and dwindling resources—all were addressed through the solutions O'Neill advocated.

However, the end of the Apollo era coincided with the rise of détente between the United States and the Soviet Union. The Cold War tensions that had motivated so much of the space race of the 1960s were dampened just as pressures on the American economy undermined the means for more grandiose space expeditions. NASA's own plans for putting humans in space were halting, hesitant, and lacked the grandeur of Apollo, while the aerospace companies that had embraced space in the 1960s were in need of new challenges. Although the Nixon administration approved the space shuttle program in 1972, some enthusiasts wondered if space exploration would ever be something that people other than a few astronaut-elites could experience.

Of course, the idea that outer space could be a tabula rasa on which more perfect chapters for humanity might be written had existed long before the earth-shattering launch of rockets from Florida's Kennedy Space Center. For instance, in the late nineteenth century, Konstantin Tsiolkovskii began thinking about how people might travel into space and what would happen to them there in the absence of gravity. To get people into space, he considered using centrifugal force and, à la Jules Verne, a giant cannon before finally focusing on a hypothetical engine that would take advantage of Newton's third law (every action causes an equal and opposite reaction). The self-taught Russian was "constantly pursued," he later said, by the "idea of traveling into space."[22] Tsiolkovskii mathematically demonstrated that space travel was indeed possible, work that helped propel the idea beyond speculation. In 1903, for example, he wrote a seminal work that outlined how a liquid-fueled rocket engine could propel a vessel into space. Besides mathematical proofs, his essays included design features that later engineers would employ, such as vanes set in a rocket's exhaust to help steer it and methods for cooling its combustion chamber.

But Tsiolkovskii went beyond designs and blueprints to cultivate and promote a broader vision for the social and spiritual implications of space travel. For example, Tsiolkovskii was friends with Nikolai Fedorov, an eccentric Russian philosopher who included writers such as Tolstoy and Dostoyevsky in his intellectual circle. Cosmism, an influential Russian philosophy in the 1920s, stemmed from Fedorov's writings. A set of ideas that integrated technology, fantasy, and emancipation, cosmism addressed the dual evolution of humanity and the universe. Space voyaging, according to Fedorov, was humanity's "common task." For devotees like Tsiolkovskii, space travel offered a path to immortality and perfection as well as a way to fulfill the Christian dream of physical resurrection.[23] Tsiolkovskii combined his theorizing and calculating with a long-term vision for the future. This, plus the ways in which those attracted to his ideas helped promote them, marks Tsiolkovskii as an early visioneer.

Tsiolkovskii was a member of a larger international cohort of space pioneers and promoters who, before World War Two, blended mathematical theorizing with utopian aspirations. Space was where, as Soviet rocket designer Fridrikh Tsander said, people could "construct a habitation in which living conditions would be much better than on earth."[24] Similar thoughts emerged in Weimar Germany and the United States.[25] Eventually, space enthusiasts like America's Robert H. Goddard moved from theorizing to building and testing their designs. But Goddard also interrupted his experiments with rocket engines to imagine the implications of humanity's "great migration" to space. Acknowledging that his speculations should be read "only by an optimist," Goddard imagined intergalactic ships carrying people to "all parts of the Milky Way" as the ultimate justification for human space exploration.[26]

Besides works by these early space-oriented dreamers and builders, O'Neill's avid science fiction reading provided him with a wealth of ideas he could tap. British author and futurist Arthur C. Clarke described an advanced settlement floating free in space in his 1952 book *Islands in the Sky*. Years later, Clarke's book *Rendezvous with Rama* told of humanity's first encounter with an alien spacecraft containing an entire artificial world inside its metal walls. Another of O'Neill's favorites was Robert A. Heinlein's 1966 book, *The Moon Is a Harsh Mistress*, a politically charged best seller that won several major prizes. Heinlein vividly described life in a lunar colony right down to the workings of its extraterrestrial economy. In Heinlein's action-packed book, rebel "Loonies" led by a "rational anarchist" professor successfully revolt against an authoritarian terrestrial government. Consciously or not, themes from Heinlein's libertarian-infused stories, which appealed to both 1960s-era Goldwater conservatives and counterculture leftists, resurfaced in O'Neill's advocacy for citizen-based space exploration.

Corporate-sponsored futurists offered O'Neill another perspective from which he could borrow. In 1964, *Fortune* profiled a cohort of "boggle-proof" long-range planners, which it called "wild birds." The magazine's term suggested a proclivity for unfettered thought accompanied by an expansive view from a lofty vantage point. The space race was a prime catalyst for these "blue-skyers" who were hired by corporate managers eager to know what the technological future might be like. Their exercises in long-range thinking, "where foresight merges with fantasy," reflected the golden age of futurology that emerged in the 1960s.[27]

One of the most adventuresome of these "wild birds" was Dandridge M. Cole, a middle-aged aerospace engineer who worked for General Electric's missile and space division in the 1960s. "G.E.'s way-out man" had a long-standing interest in what life off-planet would be like, and his interest in technological forecasting led to him write extensively about the future of space travel.[28] Cole was especially intrigued by the possibility of building permanent settlements on asteroids. Giant spaceships, each containing a "macrolife"—Cole's term for a unit of about ten thousand people plus the requisite amount of necessary plants, animals, raw materials, and machines—would travel around the solar system and provide a safety valve for continued population growth.[29] Cole's futuristic ideas rode the tide of optimistic "our-future-in-space" books that flooded bookstores after rockets launched Yuri Gagarin and Alan Shepard into space in 1961. Cole, unfortunately, never lived to see many of the wonders his writings anticipated, as he died suddenly in 1965. But fears of future "biodetonation"—his term for explosive population increase—presaged the warnings of best sellers like *The Population Bomb* and *Limits to Growth*.[30]

All these descriptions of space-related technologies and settlements in space provided O'Neill and his students with a deep reservoir of ideas to tap. Meanwhile, basic physics told them that getting people and materials out of the earth's "gravitational well" would require a lot of energy. Moreover, any new settlements on another planet, as scores of sci-fi writers had imagined, would come with another such gravity trap if one wanted to transport people and material back to earth. Instead, why not build free-floating settlements in space, far from the pull of the earth's gravity? Throughout the fall of 1969, this was the basic plan O'Neill and his students worked with. They did straightforward calculations to show, for instance, how a rotating steel cylinder could create artificial gravity and how much solar power it would need to function.

As the 1970s began, O'Neill continued to ponder the idea of space settlements late at night in his Princeton office or while flying to one of his particle physics experiments. Perhaps, in the process, O'Neill came across something Tsiolkovskii had written even as he labored in obscurity. "At first we inevitably have an idea, a fantasy, a fairy tale, and then comes scientific calculations," the space pioneer said. "Finally execution crowns the thought.[31] For years, O'Neill had been nurturing his dream with calculations. Now he wanted to see its realization.

Gerry's Worlds

While many before him had harbored utopia-like visions of permanent, large-scale human settlements in space, O'Neill carried this speculation into the realm of careful calculation. Drawing on his talent for large-scale engineering honed through his storage ring research and development, O'Neill distilled his ideas into sketches and equations. He began to methodically break down his hypothetical space colony (a term astronomer Carl Sagan and others later critiqued for its imperialist connotations) into its constituent parts. The sophistication of his treatments as well as the overall designs continued to evolve over the next decade as O'Neill revised his numbers. To be sure, he didn't resolve every engineering dilemma he faced. We might even imagine that he produced a "design for the design of a space colony." But the result was far more detailed and grounded in math than earlier conceptualizations.

O'Neill was, in effect, designing miniature Spaceship Earths. These space habitats were more complex and ambitious than the relatively modest-size earth-orbiting space stations, like *Skylab*, that the United States and the Soviet Union built in the early 1970s.

What O'Neill envisioned were microcosms of entire earthbound ecological systems. This demanded that he take physics, structural engineering, energy requirements, and, later, biology, ecology, and economics into account. O'Neill modeled his space habitats, therefore, using many of the same conceptual techniques Jay Forrester had deployed to study industrial and urban environments and which ecologists used to describe the behavior of ecosystems.

Like any scientist, O'Neill embarked on his calculations with a set of reasoned assumptions. For him, this meant designing his first prototype space colony, which he designated *Model I*, using only "technology of the 1970s or easy extensions of it."[32] This is not to say that all the technology or infrastructure needed to build it existed at the time. Some of the launch vehicles he imagined were still on the easels of aerospace engineers, and O'Neill later admitted having some ambivalence about relying on such unproven designs. But, likening himself to a driver trying to keep his car's wheels on an icy road, he endeavored to "try to keep each speculation within the bounds of numbers which can be calculated."[33]

O'Neill also looked to the technological changes he had seen in his own lifetime as another means of setting boundary conditions. When he was a boy, a DC-3 airliner could carry only a few dozen passengers some thousand miles. Three decades later, a Boeing 747 carried hundreds of passengers on nonstop transoceanic flights. Similarly, O'Neill considered the massive growth in computing power that he had witnessed over his twenty years as a research scientist. O'Neill extended the same sort of extrapolations to the tools of space travel. As a result, projections we might dismiss today as wildly overoptimistic appeared less so circa 1973, in the wake of a decade that saw such spectacular American and Soviet successes in space.

The first question he had to address was how to lift into orbit the thousands of tons of people and hardware needed to construct a space settlement. Here, he turned to NASA's evolving plans for the space shuttle, a vehicle whose predicted capabilities became a central factor in O'Neill's initial plans. As the head of NASA explained to a skeptical senator in 1972, each of the five shuttles the agency imagined entering service would to be able to carry some thirty-two tons into space with every launch.[34] The space agency anticipated this fleet making at least thirty flights per year at roughly $11 million per launch. This translated into a cost of about $200 to put a pound of payload into orbit. Time, of course, would prove NASA's projections unrealistic but, in the early 1970s, they provided a foundation for O'Neill's subsequent calculations.

O'Neill imagined an initial group of people, just a few hundred or so, who would be boosted into orbit, where they would assemble prefabricated living quarters and subsist on dehydrated food. At the same time, a few hundred more people would be stationed on the moon. There they would begin mining and processing the raw materials needed to build the actual free-floating space colony. Scientific experiments done during the Apollo missions, for example, had showed that lunar soil could be an abundant source of metals like aluminum as well as silica, which could be processed to make glass. Oxygen in the lunar material, meanwhile, could be combined with hydrogen brought from the earth to create water.

O'Neill proposed *Model I* as a pair of cylinders, about half a mile long, that would be made from the lunar-derived metal and glass, connected by tensioned cables, and positioned parallel to each other. Some six hundred feet in diameter, they would be slowly induced to rotate about their long axis until they were spinning about three times a minute to create an earthlike gravity. The cylinders would have alternating strips of windows and land. A collector at one end of the cylinder would gather sunlight to generate electricity while external mirrors would reflect light to the interior and its inhabitants. The angle of the mirrors could be adjusted to control the length of the average "day" and minimize people's exposure to harmful cosmic radiation. To get a sense of how sunlight could be introduced into a space habitat and maintain the semblance of a diurnal rhythm, he built small models with paper, tape, and plastic that accompanied his pencil-and-paper calculations.

O'Neill's plan was akin to what engineers and computer programmers call "bootstrapping." This refers to a small-scale process or program that paves the way for something more complex. Each successive space habitat would house the workforce and industries to build the next. In time, these settlements could become econom-

Figure 2.2 Gerard O'Neill's "two-cylinder" design for a space habitat that he dubbed Island Three. (Image by Rick Guidice, courtesy of NASA, NASA ID number AC75-1085.)

ically self-sufficient as they harvested resources from nearby asteroids and used these to build more habitats. O'Neill imagined that a single *Model I* might eventually hold some ten thousand people. This first wave of space settlers, using resources at hand with minimal external help, was key to O'Neill's thinking. It invoked the pioneer spirit he found attractive—his notes and writings often included somewhat naïve references to the American frontier experience of the nineteenth century—and suggested that, once underway, space settlements could become self-sustaining.

Bootstrapping would also help keep the project's initial costs relatively low. The price for building *Model I*, according to O'Neill's first estimates, was in the neighborhood of $30 billion. When compared with the cost of the era's other megaprojects, O'Neill's estimate didn't seem impossible. Experts predicted that costs for the space shuttle system and the trans-Alaska pipeline would be in the same ballpark, while Nixon-era projections for national energy self-sufficiency were far more lavish. With a estimated construction time of six or seven years, O'Neill thought the first *Model I* could be occupied by the late 1980s if the United States made a sufficiently robust investment. After all that had been spent to explore the moon, O'Neill claimed, it was "now time to cash in" on what scientists and engineers had learned from Apollo."[35]

O'Neill struggled with the issue of *where* his hypothetical space colonies should be built. He had already ruled out the surface of some other planet or moon, and an orbit around the earth would not provide the necessary gravitational stability. In mid-1973, he contacted another Princeton professor who was an aerospace engineer. His colleague suggested that one of the points in space named after the eighteenth-century mathematician Joseph-Louis Lagrange might be a good location. (O'Neill perhaps remembered then that Arthur C. Clarke had situated a space station in his 1961 novel *A Fall of Moondust* at a similar spot.) Two of the Lagrangian points, named L4 and L5, are relatively stable with respect to gravitational forces of the earth and moon. Located some 240,000 miles from earth, L5's gravitational stability meant that anything put there would remain there. O'Neill also appreciated that L5 would enjoy a steady supply of solar energy.[36]

People on *Model I* would live, O'Neill imagined, on the inner surface of the cylinders with their feet pointing out to space. The whole structure would form a closed ecological system where air and water were generated, equilibrated, and recirculated. Since hydrogen brought from the earth could be combined with oxygen released from lunar soil to make water, O'Neill imagined his colony as a lush place replete with streams, small lakes, and productive farm plots. Tons of rock and soil from the moon would shield people from cosmic rays and be landscaped into hills and valleys for recreation and habitation. More intensive agriculture could be done in smaller pods attached to the main cylinder where the atmosphere and temperature could be optimized for different crops. Similar areas for industrial work could be added, which would keep the main living area pollution-free and aesthetically pleasing.

O'Neill initially estimated that the first space colony would require some 500,000 tons of material, equipment, and people. No

EARTH-MOON LIBRATION POINTS

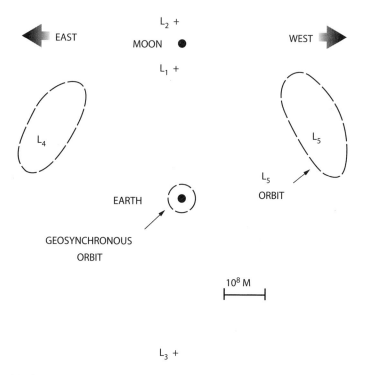

Figure 2.3 Schematic showing Lagrangian points, the gravitationally stable locations suggested for space settlements. (Redrawn from an illustration courtesy of NASA.)

matter how optimistic NASA was about the space shuttle, this was simply beyond its capabilities. Therefore, O'Neill refined his calculations and suggested that only about 2 percent of this mass should come from the earth. The lunar surface, if properly processed, could provide the vast majority of materials to build basic structures. But, even though it would take less energy to transport material from the moon's surface instead of from earth, moving tons of lunar regolith was far from trivial. When calculations showed that the plan of rockets blasting off from the moon laden with rock and soil just wasn't technically feasible, O'Neill found himself at an impasse.

To resolve this visioneering dilemma, O'Neill turned to his experience in building tools for high-energy physics. In 1956, when he started advocating storage rings, O'Neill needed a way to merge the nebulous beams of subatomic particles. His concept for a "beam line inflector" had made storage rings possible. Now, fifteen years later, he needed another breakthrough tool. What he later termed an "essential device" would allow him cut through the Gordian knot of mass, gravity, and distance and put space settlements on a firmer foundation.[37]

He found what he was looking for while leafing through the pages of *Scientific American*. In October 1973, two MIT researchers described a futuristic propulsion system for high-speed train travel. But, rather than having the train's wheels run along a track, Henry Kolm and Richard Thornton described magnetically "floating" their vehicles so that they could be propelled above metal guideways using electromagnetic forces.[38] In fact, Kolm and his MIT students had already built a small prototype that zipped along while levitating above a metal guideway four hundred feet long. Kolm had even filed patent papers for their "Magneplane." The MIT team predicted that more powerful magnets using superconductivity—a material's ability to carry a current with no energy loss—would achieve even greater speed and efficiency.

Given O'Neill's lifelong fascination with high-tech transportation and his hard-won expertise with complex systems of vacuums, magnets, and electronics, the possibility of using electromagnetic forces to propel objects immediately appealed to him. Inspired by the *Scientific American* article, O'Neill began to consider how a similar device might transport rock and soil from the lunar surface. As he worked out the technical details, he was again following in the footsteps of his favorite science fiction writers. For example, Heinlein described lunar catapults in *The Man Who Sold the Moon* and *The Moon Is a Harsh Mistress*. Moreover, a nonfiction piece by Arthur C. Clarke predicted an electromagnetic launcher on the moon, serviced by a "self-supporting lunar colony," that would propel lunar rock into space to fuel Mars-bound craft. Clarke even noted how Westinghouse had already built an "Electropult" to help boost the takeoff speed of small military aircraft.[39]

Over the next few months, O'Neill sketched out a design for what he called a "mass driver." O'Neill envisioned it using small, magnetically levitated metal buckets on the moon's surface. An engine, powered by solar energy, would accelerate the buckets above a long and gently rising metal guide track along which superconducting coils were spaced at regular intervals. Sequential "firing" of these coils could then accelerate the buckets to the moon's escape velocity. If, at that point, a bucket's speed was carefully reduced, it would fling its payload of compacted lunar soil off into space before circulating back to be filled again. With each bucket "launching" some ten pounds every few seconds, a sufficiently large mass driver could, over several years, transport hundreds of thousands of tons of material to a space-based processing site.

Given its relative simplicity—the underlying physics dated to the late nineteenth century, and there were few moving parts other than the buckets—O'Neill saw the mass driver as an elegant solution. It violated no laws of physics and, after O'Neill carefully worked out its power requirements, dimensions, and performance, he concluded that it could be built with existing or soon-to-be available equipment. But, like the rest of his designs for space settlements, the feasibility of O'Neill's mass driver was predicated on a series of optimistic assumptions and extrapolations: a lunar outpost could be established, physicists would see continued progress in improving the capabilities of superconducting wires, and so forth. And all this, of course, rested on another, broader set of assumptions about economics, the accuracy of NASA's long-term projections, and sufficient public support. As O'Neill and other visioneers all discovered, just having a sound set of calculations, some inspiring drawings, and a vision for the future wasn't enough.

Going Public

By mid-1972, O'Neill had filled a stack of notepads with numbers, graphs, and sketches. These, in turn, informed the very rough manuscript he had written, something he called simply "The Colonization of Space." But what pushed him to go beyond late-night calculations and doodling in airports and "take it to the people," as he later phrased it, was the global uproar caused by the *Limits* report as journalists and academics framed the future as something to fear.

Although eager to get colleagues' feedback and see his ideas reach a wider audience, O'Neill instead started accumulating rejection slips. *Scientific American* turned down his manuscript outright, and the *Atlantic Monthly* said his ideas raised more questions than they answered. He next tried the journal *Science*, the flagship publication that the American Association for the Advancement of Science publishes weekly. The response O'Neill received from *Science*'s anonymous reviewers was typical of the problem visioneers often encounter when they try to promote their future-oriented engineering. The referees found O'Neill's manuscript "significant, technically sound . . . authoritative." However, both readers found themselves pulled between the technical feasibility of O'Neill's ideas and broader questions of politics, economics, and societal needs. Despite the space program's impressive technical accomplishments, one referee asked "Whoever said that Project Apollo was economical or a reasonable thing to do?"[40] Just because a space colony *could* be built didn't mean that it *should* be built. O'Neill added another rejection slip to his pile.

Somewhat at a loss, the physicist sought out others to talk with about what he called his "new, delicate and easily killed idea."[41] Many people before O'Neill, of course, had been seduced by space, drawn into its dreamworld of calculation and speculation. Ironically, he discovered one of his best sources of insight and criticism in his hometown. Freeman J. Dyson, a British physicist who was a few years older than O'Neill, had been interested in radical approaches to space exploration for more than a decade. The two physicists, who were well aware of each other's physics research, began to exchange ideas and speculations about the technological future.

Dyson's sterling reputation as a physicist was based on his theoretical work on the fundamentals of how light and matter interact at the quantum level. After coming to the United States in 1947, he studied physics at Cornell, O'Neill's alma mater. A year later, physicist Robert Oppenheimer recruited Dyson to the Institute for Advanced Study in Princeton. But, after the launch of *Sputnik 1* in 1957, the polymathic Dyson took a leave of absence from theoretical physics to work on Project Orion—this was a design study for a proposed interplanetary craft that would be propelled forward by nuclear explosions.[42] As bizarre as such an effort might seem today, military and NASA funding helped the Orion team build a small (nonnuclear) test version of their space exploration machine. However, the signing of the Limited Test Ban Treaty in 1963 and dwindling air force interest contributed to Orion's cancellation in 1965 even though Dyson and others had contributed years of serious design and engineering work.[43]

Orion's cancellation taught Dyson an important lesson. He learned that the U.S. space agency was not some monolithic entity with all its branches and research centers pulling in the same direction. Instead, like all large organizations, it was riven by different, sometimes opposing, senses of where it was going and how to get there. At the same time, external forces such as Washington politics and the needs of aerospace contractors also pulled on it. As Dyson later wrote, there was the "real NASA," which was "inherently conservative, dedicated to preserving existing programs," and there was the "paper NASA," which was "adventurous on paper" but penurious in terms of resources and power. Dyson, and later O'Neill, learned the hard way that the "real NASA" almost always trumped the paper version.[44]

Still, perhaps more than anyone else O'Neill had met before, Dyson understood the pull that outer space could exert on a scientist's imagination. Even as he worked on Orion, Dyson wrote a "Space-Traveler's Manifesto." In it he identified space as the last remaining place where small groups of people might still "escape from their neighbors and from their governments, to go and live as they please."[45] This reason, which seemed even more imperative after dire reports of the planet's future appeared, made Dyson believe that human space exploration remained vital. In the spring of 1972, as the furor over *Limits* neared a crescendo, the two physicists debated technical points about space settlements via a series of letters. However, Dyson admitted the design details mattered less to him. The main thing was "to get people to think seriously" about the possibilities for humanity's future move into space.[46]

Dyson's discussions with O'Neill came at a fortuitous time. In May 1972 London's Birkbeck College welcomed Dyson as he gave the school's annual Bernal Lecture. The lecture series was named after John Desmond Bernal, an Irish crystallographer and molecular biologist. As he considered his lecture topic, Dyson decided to revisit a visionary book about the future Bernal had written early in his career.

Published in 1929, *The World, the Flesh, and the Devil* was Bernal's succinct exploration of how radical new technologies could help society confront what he called the "three enemies of the rational soul."[47] Bernal's first and foremost foe was the World. To transcend the limits of terrestrial resources and the unpredictability of the planet's environment, he proposed that human habitation expand into the cosmos. To aid this migration, future engineers would craft "new molecular materials."[48] In permanent free-floating settlements, "free communication and voluntary associations of interested persons" would prevail, liberating people from earthly traditions and values.

To thrive in these new environments, people would be obliged to "interfere in a highly unnatural manner" with the Flesh. For Bernal, this first meant radical surgery, the replacement of organs and tissues with mechanical substitutes, and then the eventual modification of people's genetic material. These new and improved humans would learn to reproduce in new ways—Bernal was a devoted libertine—and eventually seek a form of immortality by preserving their memories with electronics and machines. However, the Devil, which Bernal defined as our "desires and fears . . . imaginations and stupidities," remained a treacherous foe for people who wanted to actively engineer their future.

In his lecture, Dyson concluded Bernal had been quite prescient.[49] People had gone into space, artificial organ transplants were now possible, molecular biology was an established field, and knowledge about the workings of the human brain had grown greatly. Even when it came to the Devil, Dyson found Bernal on the mark. In 1972, just as in 1929, there was a "highly vocal and well-organized opposition" to the "further growth of technology" as irrational "social prophets" depicted it as a "destructive rather than liberating force."[50] The difference was that whereas Bernal saw the "defeat of the Devil" happening via a combination of socialism and psychology, Dyson imagined a future in which the Devil no longer had the whip hand, and this would only happen when people faced the threat of limits with courage and foresight, not fear.

Dyson also used Bernal's slim book as a springboard for his own recommendations. He suggested that new technologies must be harnessed as allies in the struggle against resource scarcity, overpopulation, and other planetary ills. For example, biological engineering could design novel organisms that could convert "wastes efficiently into usable solids and pure water." Inorganic "self-reproducing machinery," analogous to coral and oysters, could collect minerals from the ocean and be another ally in the "attack on Bernal's three enemies." Besides ameliorating life on earth, radical bioengineering could also enable the colonization of space. New life-forms such as "Big Trees," as Dyson called them, could be engineered to grow on comets that had ample amounts of water and nitrogen. The result might be the "greening of the Galaxy" as settlers transformed space into habitable oases.[51]

The success of the Apollo era helped shape Dyson's thinking about the technological future. So did his experiences as a father. His trials as a parent became public knowledge in 1978 with the appearance of popular book written by Kenneth Brower, son of Sierra Club founder David Brower. *The Starship and the Canoe* chronicled the tensions between the famous physicist, who worked at the intersection of science and the military (Freeman Dyson regularly advised on national security issues), and his rebellious son. The book's narrative eloquently juxtaposed Freeman's desires to build nuclear-powered spacecraft and live in space with those of his son, George, who was building a giant oceangoing kayak while residing in a towering Douglas fir tree in British Columbia.

The elder Dyson's own writings about space exploration also reflected the gap between one generation of dreamers and another. In them, it's clear that the physicist was writing, if not to, then about, his son (as well others of his son's generation who were unenthused about modern technology). Space could be a refuge for "rebels and outlaws . . . recalcitrant teenagers escaping from their parents . . . beyond the reach of snooping policemen and bureaucrats." This was, of course, a view of the future that contrasted with scenarios of managed "spaceman economies" and the limits of Spaceship Earth. By offering an ideal frontier for "angry young men and rebels," space exploration might also give more "timid and law-abiding citizens" a chance to "live together in peace."[52]

Dyson and O'Neill agreed on the general merits of human settlements in space, but they differed as to what this humanization of space might look like. "Your standardized prefab communities have a bureaucratic quality which repels me," Dyson told his Princeton neighbor. Although "my pipe dreams are different from yours," Dyson noted, the planet needs "many dreamers and many dreams" from which "fate will make her choice."[53] O'Neill genially replied that he, too, hoped that space settlements would be a way to explore new frontiers and argued that space communities need not be standardized. "Taormina and Stow-on-the-Wold [small tourist towns in Sicily and England respectively] are both attractive," O'Neill said, "but very different and about equally artificial!"[54] Despite their differing views about how the humanization of space might happen, Dyson recalled that his Princeton neighbor got him seriously interested in space exploration again after the disappointment of Project Orion. "O'Neill revived me," he said.[55]

In the fall of 1972, frustrated by his inability to get his manuscript published, O'Neill opted for a more grassroots approach. Here, he was following advice given by supportive colleagues who advised him to "remember [Robert] Goddard and don't get discouraged."[56] His first stop was Hampshire College in Massachusetts. The invitation was arranged by Brian O'Leary, a teacher at Hampshire and former astronaut who had been part of O'Neill's NASA training cohort. By all accounts, O'Neill's visit was a huge success, and scores of students engaged him about his ideas. As he continued to travel for his research, O'Neill gave subsequent presentations at Caltech, Stanford, and Berkeley, where give-and-take sessions with audiences pushed him to address unresolved technical questions.[57]

Finally, in mid-1973, after more than two years of revisions and rejections, O'Neill received positive news about his manuscript. Harold Davis, the editor of *Physics Today* (*PT*), accepted it for the journal. Published by the American Physical Society and sent out to tens of thousands of scientists every month, *PT* blended news and commentary on the international physics community with articles, written by experts, on particular scientific topics. In the late 1960s, *PT*, like many science journals, had started to focus more on the nexus of science and society, and O'Neill's ideas fit this editorial goal.

Now assured that his ideas would soon reach a wider audience, O'Neill started sending occasional newsletters to students and colleagues, many of whom had attended his campus presentations. His first such missive, which he wrote on Christmas Eve in 1973, made it clear how the furor over *Limits* and his enchantment with new frontiers remained central to his thinking. "The sense of new horizons, new opportunities, of a freedom from limits, is a heady brew," he told the hundred or so people on his initial mailing list. He also announced his intent to convene a small workshop on space colonization at Princeton the following spring.

To do this, O'Neill needed money. He started first by talking to people at private foundations. After several phone calls, however, he was getting nowhere. A colleague suggested that he try the Point Foundation, a small nonprofit organization in San Francisco started by Stewart Brand. In the early 1970s, Point had been flush with money from the immensely successful *Whole Earth Catalog* enterprise. The foundation's eclectic philanthropic focus included social justice and a pragmatic approach to environmentalism. Although O'Neill's vision was not typical of the small-scale technologies presented in books like E. F. Schumacher's 1973 classic, *Small Is Beautiful*, it resonated with Michael Phillips, Point's director. Willing to take risks and pursue unconventional, if not eccentric, new ideas, Phillips recognized that O'Neill's mega-scale concept comported with Point's general interest in technology, environmentalism, and experimental communities.[58]

Unfortunately, by 1974, the Point Foundation was running low on funds, and the best Phillips could offer O'Neill was $600. This was several orders of magnitude less than the budgets of the big physics projects he had worked on. But it was enough. While not his professed intent, having a connection to Brand and the Point Foundation gave O'Neill, a physicist and unabashed fan of large-scale technology, a modicum of credibility among the communalists and back-to-the-land folks who read *Whole Earth*. Phillips shrewdly proposed that Point's modest award go not to O'Neill but to Princeton University. Making it a formal grant would force "the Establishment [to] recognize the existence of your work by filling out forms and generating a lot of red tape . . . the only reality that is understood."[59]

On May 9, 1974, O'Neill met with a select group of people who had an especial interest in the technical and social aspects of space colonization. Freeman Dyson attended as did representatives from NASA and several university students, including Eric Drexler, who trekked down from MIT. The next day over a hundred people came for an open discussion of topics ranging from detailed descriptions of space shuttle designs and methods for refining lunar materials to speculations about the social implications of permanent space settlements.[60]

Phillips's proviso to make Point's modest award a formal grant to O'Neill's school had an unexpected effect. Up to this stage, O'Neill had been reticent to discuss his ideas with reporters, preferring to focus on design details. But as the date for the "officially Princetonian" gathering approached, the school's news bureau sent out a press release. As a result, O'Neill's small gathering drew the attention of reporters, including science writer Walter Sullivan from the *New York Times*.[61] Sullivan was renowned for covering everything from the first satellite launches of the Space Age to the search for extraterrestrial intelligence and nuclear physics. The talks Sullivan heard from the people gathered in Princeton resonated with the tales of bold explorations he had covered as a young reporter.

The following Monday morning, as Gerry and Tasha O'Neill were packing for a summer of research and travel in California, the *New York Times* published Sullivan's article on its front page. The headline announced that human colonies in space were "Hailed by Scientists as Feasible Now."[62] Other reporters, includ-

physics today

SEPTEMBER 1974

Colonies in space

Figure 2.4 Cover of the September 1974 issue of *Physics Today*, one of the first publications showcasing Gerard O'Neill's visioneering. The painting, by Walter Zawojski, an artist at the Stanford Linear Accelerator Center, depicts one of O'Neill's cylindrical designs for a space habitat. (Image reprinted with permission from *Physics Today*, copyright 1974, American Institute of Physics.)

ing ones from *Time* and the BBC, picked up the story, and a small avalanche of articles and television stories about O'Neill and space settlements accumulated over the summer. After the O'Neills returned to Princeton for the fall semester, his *Physics Today* article appeared as well. Like Sullivan's story, it too earned feature status—*PT*'s cover gave a visual interpretation of O'Neill's idea, depicting a small shuttle craft approaching a cylindrical settlement, slim and spartan, floating in space.[63] After five years of writing and revising, the ideas O'Neill believed in were starting to reach citizens and scientists alike.

Space as a Place

More than anything, what sparked public and media curiosity in O'Neill's ideas was his depiction of space not as a government-run *program*, but as a *place*. This critical shift in perspective, of seeing space as real estate for potential habitation and manufacturing, was essential in motivating the first wave of people excited by his visioneering. It is nearly impossible, though, to fully understand O'Neill's ideas without seeing them as concepts that fit into broader narratives about technology that were already tightly woven into American culture by 1970.

O'Neill's plan for settlements in space was an amalgam of engineering-oriented conceptualizations coupled with established ideas about how technology had helped create modern American society. As a hybridized and evolving vision, O'Neill's advocacy of space settlements blended his fascination with large-scale engineering and his belief that technology could offset cultural stagnation. He also believed his plan offered pragmatic solutions to social issues. For example, the humanization of space could help avoid crises—energy shortages, habitat destruction, resource depletion, and overpopulation—that loomed on the horizon. And it could benefit all people, not just Americans. "We *can* colonize space," he wrote, "and do so without robbing or harming anyone and without polluting anything."[64] Like the commune builders of the late 1960s, O'Neill hoped his space settlements would become sites of racial harmony and offer expanded roles for women and minorities.[65] With their infusion of a certain "counterculture libertarianism" and an emphasis on environmentalism and equal opportunities for adventuresome individuals, O'Neill's ideas defied easy categorization.

Some critics would take O'Neill to task for proposing an impractical, politically naïve, and hopelessly expensive megaproject. However, labeling it as flawed from the outset imposes a judgment made with the benefit of hindsight. In the context of the era's other extravagant schemes—Apollo, the shuttle, the MX missile system, the Alaska pipeline, and even Project Orion—O'Neill's ideas ap-

pear less grandiose. His idea of space colonies responded to prevailing pessimism about technology and profound concerns about
the deteriorating environment while embracing the can-do confidence that O'Neill saw as a legacy of the Apollo era. Also, one
must recall that, given the successes of the U.S. and Soviet space
programs, O'Neill's idea displayed a certain internal logic derived
from a sense of technological momentum.

O'Neill would repeatedly deny that space colonies were a panacea for all of society's ills. "There are no Utopias," he insisted.[66]
Nonetheless, his early articles and his best-selling book, *The High
Frontier*, can be situated within a larger literary tradition of technological utopias.[67] But where previous technological utopians offered only descriptive "literary blueprints" for change, O'Neill's
visioneering offered something more—engineering studies, detailed designs, machinery schematics, and cost analyses.

Taking a cue from Freeman Dyson, O'Neill imagined space
communities offering places for "free, diverse social experimentation" aided by a "new technical methodology."[68] His depiction of
detached, self-sufficient communities was reminiscent of earlier
utopian settlements like Oneida in upstate New York or, more recently, Le Corbusier's *Unité d'Habitation* in Marseille or Paolo
Soleri's Arcosanti outside Phoenix, Arizona.[69] In fact, the ideal of a
shared, purposeful high-tech community is exactly what appealed
to many of O'Neill's early supporters.

Like earlier technological utopias, O'Neill's future-oriented designs proffered new social possibilities at what historians would
come to see as a turning point—militarily, politically, culturally, and
economically—in the United States' history. For those captivated
by O'Neill's radical vision, the idea of living and working in space
offered a sense of hope and purpose. O'Neill never made a strong
claim that space colonies *would* result in a perfect society or solve
humanity's problems of environmental degradation, racism, and so
forth. He did, however, believe that the permanent expansion of
human habitation into space could at least provide an opportunity
to ameliorate social, environmental, and economic anxieties.

Early twentieth-century technology enthusiasts had lauded the
construction of giant dams, water systems, and canals as a master-

ing of nature. Likewise, O'Neill's vision would require the reordering and remaking of nature.[70] His ideas reflected what historian Leo Marx famously termed the "machine in the garden" as nineteenth-century industry and the American pastoral ideal coexisted.[71] In O'Neill's case, however, the garden itself was *in* the machine. Technosystems and ecosystems merged to create a new and harmonious relationship. And where nineteenth-century utopians like Edward Bellamy saw power provided not by smoke-belching coal furnaces but by quiet electric turbines, O'Neill imagined that solar power, a technology improved by the space program and approved by environmentalists, would provide clean energy. And with plentiful, clean energy, space-based settlements would pave the way for space-based manufacturing and move polluting industries away from the earth. This offered a compelling argument in the mid-1970s. It is telling that the focus—indeed, the name—of the regular conferences O'Neill helped organize shifted from space colonization to space-based manufacturing after the initial 1974 Princeton meeting.

O'Neill was not the only future-minded person with a vision of a new economic order and the reconfiguration of American industry. In 1973, Harvard sociologist Daniel Bell published *The Coming of Post-Industrial Society*, a best seller which, in keeping with the era's preoccupation with prognostication, he subtitled *A Venture in Social Forecasting*. Along with management guru Peter Drucker and pop futurist Alvin Toffler, Bell declared that a major technological and societal shift was underway. The transition to a postindustrial society meant "a changeover from a goods-producing society to an information or knowledge society."[72] The traditional, centralized industries that churned out cars, dishwashers, and so forth would be eclipsed by more flexible forms of production oriented instead around information and services—what Toffler later billed as a "third wave." These futurists predicted a new industrial revolution that would catalyze widespread social, economic, and political upheavals. O'Neill differed from people such as Bell or Toffler, of course, in that he added considerable engineering and design work to advance toward the particular future he had in mind.

O'Neill's visioneering also resonated with political ideas promulgated by countercultural thinkers like Murray Bookchin. An environmentally conscious social philosopher who began his career as an anarchist, Bookchin described a "post-scarcity economy" that could reconcile Americans' enthusiasm for technology with their environmental concerns. (Here, postscarcity referred to the ability of industrial society to meet its citizens' needs.) In a 1965 essay called "Towards a Liberatory Technology," Bookchin rebuffed the "blanket rejection of technology," insisting that it could still provide abundant resources while avoiding both "centralized economic control" and the destructive tendencies of unbridled capitalism.[73] O'Neill's space settlements were based on similar imaginings of technology's liberating potential for creating abundance, ideas that, as we'll see, early enthusiasts for nanotechnology like Eric Drexler embraced in the 1980s.

The ghost of a famous historian, however, haunted O'Neill's imaginary space settlements. In 1893, Frederick Jackson Turner argued that, from the first European settlements in the New World until his own time, the promise of "free land" had been a "consolidating experience" that had shaped American society and encouraged democracy and individualism. Westward expansion to new frontiers, Turner concluded, had offered the American people an essential "gate of escape" where cultural reinvigoration could occur free of the "scorn of older society ... its restraints and its ideas."[74] Later historians discredited Turner's argument roundly. The Western frontier was hardly an "empty" place, and colonists from Europe were certainly not economically and technologically self-sufficient. Nevertheless, the simple causality and emotional appeal of a beckoning, rejuvenating, invigorating frontier made Turner's thesis hugely influential on many people, O'Neill included, who supported a manned space program.[75] O'Neill, like Freeman Dyson, projected ideals of societal regeneration and a sense of manifest destiny onto space as the new frontier. An essay O'Neill wrote toward the end of his life made explicit the analogies he saw between his own ideas and the experiences of American settlers. "We can look," O'Neill said, "to our own nation's history as a guide to our future ... first will come the prospectors, then the miners, then the builders."[76] But the future didn't have to repeat the past in all ways. Settling a new, off-planet frontier could, for instance, avoid reprehensible aspects of traditional colonialism and permit expansion "without shooting any Indians."[77]

Looking beyond his somewhat idealized and naïve images of the frontier, we find O'Neill's visioneering tapping into older origins stories that dated back to the founding of the republic. Since the late eighteenth century, national narratives of the United States gave a privileged role to what Frederick Jackson Turner had called Americans' "masterful grasp of material things." For colonial settlers, these included the railroad, the ax, the log cabin.[78] For O'Neill and hopeful space colonists, this now meant the rocket, the mass driver, and the rotating cylinder. In both cases, what resulted was an imagining of endless abundance and personal freedom coupled with laissez-faire thinking.

O'Neill balanced his visioneering between centuries-old concepts of frontiers and utopias, profound social and environmental concerns, and the technological realities of his time. America's first colonists imagined themselves establishing "a second creation built in harmony with God's first creation."[79] In this narrative, the natural world remained unfinished, a space waiting for Anglo-Saxon settlers to intervene and improve it. Just as the settlement of the American West imagined the imposition of order on a wild and "empty" landscape, so did O'Neill and his supporters see space as a truly uninhabited medium in which new self-sufficient and independent communities could take root. At the same time, O'Neill's humanization of space posited a new technologically mediated narrative of national, even planetary, regeneration—a third creation, perhaps—in which average citizens, not just specially trained astronauts, could take part. The man from Speculator now waited to see how the public and policy makers would respond.

CHAPTER 3

Building Castles in the Sky

|||||||||||||||||||||||||||||||||||||||||||||||||||||||||||||||||||||||||||||||||||||||||||||||||||||||||||||||||||||||||||||||||||||||||||||||||||||||||||||||||||||||

We want our Utopia now—and we're going to try our hands at it.
—Sinclair Lewis, *Main Street*, 1920

While searching through papers related to Gerard O'Neill's career, I found an intriguing photograph. Taken on January 25, 1978, it shows O'Neill testifying before a congressional committee. Books and articles outlining his visioneering ideas are stacked in front of him. O'Neill himself looks distracted or perhaps just tired. He had been traveling a great deal to promote his ideas. Seated next to him is Barbara Marx Hubbard, a wealthy social activist and space enthusiast who advocated what she called humanity's "conscious evolution." Behind O'Neill, his eyes turned toward someone in the gallery, is Eric Drexler. When the photo was taken, Drexler, a recent graduate from MIT, was already a seasoned advocate of O'Neill's ideas.

O'Neill's expansive vision for the humanization of space catalyzed a small-scale social movement. As such, it had some features of other social movements: a shared vision for change, political activism, formal organizations, and informal networks of like-minded supporters. This photo, taken near the peak of public visibility for O'Neill's ideas, testifies to the diversity of people who coalesced around them.

Despite their varied backgrounds, this cohort rejected, as a rule, scenarios of a future with proscribed limits—economic, environmental, or otherwise. Instead they voiced their support for the opening of new technological frontiers. These enthusiasts, some of whom had protested against the Vietnam War or rallied for women's rights and environmental causes, saw the humanization of

Figure 3.1 A pensive Gerard O'Neill testifying before Congress in January 1978, at the "Future Space Programs" hearings. Sitting to his left is futurist Barbara Marx Hubbard. Behind O'Neill, in the black turtleneck and looking off to the side, is K. Eric Drexler. (Image courtesy of Tasha O'Neill.)

space as a potentially new and powerful form of social action, one that might benefit society as well as themselves.

The "humanization of space" also reflected enthusiasts' desire to escape, reinvent, or start over, a theme central to influential films like *Bonnie and Clyde* (1967), *Butch Cassidy and the Sundance Kid* (1969), and *McCabe and Mrs. Miller* (1971). All of these portrayed outlaw heroes pursuing unconventional lifestyles while trying to escape the Establishment's oppression. The young adults who gravitated to O'Neill's ideas could easily imagine themselves playing out similar scenarios against the pseudo-Western backdrop of space, the final frontier.

In the mid-1970s, O'Neill's ideas and designs for human settlements in space began to secure a beachhead in American popular culture. This chapter shows how O'Neill extended the engineering-oriented foundation for his visioneering to a wider base of enthusiasts. Through conferences, workshops, and the accretion of new ideas, O'Neill continued to describe a future in which space-based settlements remained plausible, at least in technical terms, and desirable. At the same time, his "humanization of space" idea mutated as journalists, politicians, writers, college students, and counterculture figures embraced or opposed it. This inherently messy process reflected a decade marked by social confusion, political realignment, and economic uncertainty. And, as happened with many of the other social movements from the 1960s and '70s, enthusiasm for the humanization of space produced both strange alliances and unexpected outcomes.

Students for Lagrangia

O'Neill wanted to stimulate discussion and debate. In addition to the professional researchers he was familiar with, he was eager to hear from "people with a range of technical and artistic talents . . . people who claim no special talent beyond the ability to work hard for a worthwhile goal."[1] After the publication of his *Physics Today* article, scores of letters and phone calls from other scientists, curious citizens, and cranks started to arrive at his Princeton office.[2] Some people wanted more information, some wanted to debate technical points, and others just wanted to express support (or tell O'Neill why he was wrong). The volume of correspondence grew such that in his semiregular newsletters O'Neill apologized en masse to correspondents he had not replied to and asked for volunteers to assist him.[3] O'Neill meanwhile maintained a demanding lecture and travel schedule, regularly giving talks to NASA teams, aerospace companies, and campus groups.

Proselytizing combined with steady media coverage helped make O'Neill a minor celebrity. Articles in the *New York Times*, *Harper's*, and *Time* popularized the "humanization of space" as did niche publications like *Penthouse*, *Popular Mechanics*, and the *Futurist*. O'Neill's local radio show interviews were complemented by his appearances on national television shows—*Today*, *The Tonight Show*, *60 Minutes*, *The Merv Griffin Show*—which millions of Americans watched. An even bigger boost of publicity came when America celebrated its bicentennial. *National Geographic* published a lavishly illustrated feature on the "first colony in space" penned by Isaac Asimov and shot through with classic "final frontier" themes.[4] One NASA official recalled initially receiving just a few dozen letters from citizens curious about space settlements. After the *National Geographic* article appeared, his office received several thousand more.[5] Intrigued people wanted more information.

The approach that reporters and writers used to address O'Neill's ideas varied widely. *Science News*, a U.S. publication, and the *New Scientist*, based in London, both gave technically accurate accounts about how, with enough money and time, thousands of people might one day take up residence at "Lagrangia."[6] Talk show hosts took a variety of tacks, from "gee whiz" predictions about future societies in space to depictions of O'Neill as a scientist-rebel trying to think outside NASA's bureaucratic box. The *New Yorker*, not surprisingly, assumed an arch tone. A short essay depicted the Princeton physicist and his supporters as escapists who sought to flee "the jostle and noise of Earth" for a refuge it lampooned as "Exorbia." At a time when "white flight" remained a major social concern, criticisms that O'Neill's scheme was a high-tech variation on suburbanization were potent.[7]

Nonetheless, O'Neill's humanization of space proved especially attractive to college students, especially those in the sciences or engineering. (O'Neill himself was surprised to find that "young Maoists" were enthusiastic supporters as well.)[8] This student interest, especially from the counterculture, might seem odd given that O'Neill was a tenured physics professor with a background in big-bucks Establishment science projects at a quintessential Establishment school. However, his popularity reflected an era of iconic intellectuals, from Marshall McLuhan and Noam Chomsky to Immanuel Velikovsky and Thomas Kuhn, whose best-selling books and radical ideas generated dorm-room debates and helped fill lecture halls.

Eric Drexler was one of the first students to feel the pull. Like many technology enthusiasts drawn to O'Neill's vision, Drexler rejected the assumption that people should consider the planet as a closed system with inherent limits. Abundant solar energy flooded the planet, and relatively nearby bodies such as asteroids and the moon could provide boundless material resources. In 1973, when he applied for college, Drexler set his sights on MIT. "I would like to study what needs to be done," he wrote in his college application, "for the founding of a self-sufficient [space] colony capable of growth and find ways of making such a project politically feasible." For the next several years, this ambition drove the young and idealistic polymath.[9]

Drexler arrived at MIT at a curious time in that school's history. In 1969, some of MIT's faculty staged a widely publicized one-day strike and teach-in to protest abuses of science and technology and their school's participation in defense-related work.[10] When Drexler started his studies in 1973, the institutional reforms that this and other demonstrations had prompted were still underway. For example, Drexler declared interdisciplinary science as his major. MIT had established this experimental program in 1971 to allow students like Drexler to pursue coursework spanning several fields.[11]

After settling into student life at MIT, Drexler discussed his interest in space exploration with his professors. One of them, Philip Morrison, was sympathetic to radical ideas with potentially revolutionary outcomes. As a young physicist, Morrison had worked on the Manhattan Project and had witnessed the first explosion of an atomic bomb at the Trinity test in New Mexico. He later coauthored a seminal paper that helped create the Search for Extraterrestrial Intelligence project. Drexler took Morrison's advice to travel to Princeton and meet with O'Neill. Soon after they talked, the Princeton physicist offered to hire Drexler as a research assistant for the summer of 1974.

Just as he had done for his first Princeton conference, O'Neill relied on an unconventional patron to help pay his new assistant. Barbara Marx Hubbard was an heiress—the money came from a toy company started by her father—and a futurist of sorts. Like Pierre Teilhard de Chardin, a controversial French priest and philosopher Hubbard admired, she believed humans and the universe were evolving together, perhaps toward what de Chardin termed the "Omega Point." When she witnessed the *Apollo 11* launch,

Hubbard recalled how she felt herself "rising in space . . . the words 'freedom, freedom, freedom' pounding in my head" as the mighty Saturn V rocket blasted through the atmosphere.[12] As Hubbard developed her own "philosophy of hope," she worried that people would be "incarcerated in a closed system of increasing control and depletion of resources."[13] Outer space, for Hubbard, meant both personal liberation and a way for society to escape the planet's confines.

In 1970, with her artist-philosopher husband, Hubbard formed the Committee for the Future, a small group of "space activists, dreamers, and misfits." Its first major ambition was a "citizen-sponsored lunar mission." Hubbard imagined the expedition would use surplus Apollo program hardware and support itself by selling media rights along with pieces of moon rocks it returned.[14] Hubbard presented the idea to representatives of Congress, UNICEF, and NASA, her access enabled by her family's money and political connections. She was no stranger to controversy; her sister was married to Daniel Ellsberg, the military analyst who had famously leaked the "Pentagon papers."

When the Committee for the Future's quixotic endeavor failed, Hubbard looked in other directions. With a $100,000 infusion of family funds, she began organizing "synergistic convergences," dubbed SYNCONS, at colleges and conference venues around the country.[15] These gatherings blended space and technology-oriented enthusiasm with the New Age mysticism that was becoming increasingly fashionable as many Americans experienced a new Great Awakening amidst a burst of spiritual and religious experimentation.[16]

Around 1972, Hubbard encountered an essay called "The Extraterrestrial Imperative" written by German rocket engineer Krafft Ehricke.[17] Calling him the "Darwin of technology," she agreed with his argument that expansion into space was both inevitable and necessary. Ehricke's essay and *Limits to Growth* reflected a tension she saw between a future of expansion versus one of stagnation. In response, Hubbard started a small research group devoted to questioning the *Limits* thesis, considering the possibilities of postindustrial society, and promoting the exploration of "inner and outer space." To explore these ideas further, Hubbard invited O'Neill to Greystone, her mansion in Washington. Impressed with what she heard, Hubbard agreed to help fund Drexler's summer research on the chemical processing of lunar soil.

When he returned to MIT that fall, Drexler shared his enthusiasm for the humanization of space with other students. He organized a student seminar on space colonization and started a small but enthusiastic "Space Habitat Study Group." Interest at MIT in O'Neill's ideas eventually made its way into the school's curriculum. In the fall of 1975, the Department of Aeronautics and Astronautics offered a new undergraduate course on "space systems engineering" that explored some of O'Neill's concepts. A more intensive course, led by John F. McCarthy, director of MIT's Center for Space Research, followed. For an entire semester, undergraduate and graduate students studied the engineering needed to build a small, industrial space settlement. Students called their habitat the "MIT can" and designed it to support a few thousand people.[18] The MIT study presaged several other efforts at campuses around the country that adapted O'Neill's visioneering for pedagogical purposes.[19]

Drexler was part of a growing group of college students, many of them majoring in science or engineering, attracted to the humanization-of-space movement. What drew some was the technical challenges that implementing O'Neill's plan posed. Others responded to O'Neill's general rejection of the *Limits* thesis coupled with an optimistic sense of the possibilities that a large-scale, permanent human presence in space might afford in their lifetimes.

Mark S. Miller, a Yale undergraduate studying computer science, was one of these students who liked what O'Neill had to say. As a teen, Miller had discussed politics and economics with a favorite uncle who was an engineer. Heinlein's *The Moon Is a Harsh Mistress* stirred his interest in libertarian themes, and he even changed the *S* in his middle name to *$* for a short time. When Miller first encountered O'Neill's ideas, his response was "Of course! It made such complete sense." Now the humanization of space was "a question of logistics" and of making it "economically sensible." In the mid-1970s, Drexler met Miller, and the two formed a long-lasting friendship. "It just became very clear that the earth is just a starting point," Miller said, recalling their shared excitement. "It was the feeling of being at the beginning of a grand adventure."[20]

Harvard student Mark M. Hopkins also believed that humanity's expansion into space could solve a host of societal and environmental problems. The financial angle especially interested him. After getting his undergraduate degree from Caltech in 1971, Hopkins studied economics at Harvard. He also helped found the Harvard-Radcliffe Committee for a Space Economy. Hopkins wondered whether space settlements could be economically viable and, after reading O'Neill's *Physics Today* article, he phoned the physicist and bluntly asked "How are you going to pay for this?"[21]

Pleased to have more people probe his plan for weaknesses, O'Neill invited Hopkins, Miller, and Drexler to attend the next Princeton conference on space settlements. More formal than the first event, with many more technical presentations, the May 1975 gathering continued to generate media attention. Reporters from major American and European publications came to hear talks oriented around themes such as industrial production in space, "new options for self-governance in space habitats," and "diversity and lifestyle enrichment."[22] More than three decades later, Hopkins still remembered the excitement the meeting generated. "You could just feel it," he said. "I've been to very few conferences like that."[23]

After the Princeton meeting ended, Hopkins and Drexler headed to California, where they joined a summer study for the "design of a system for the colonization of space." This was based at the Ames Research Center, NASA's facility in the Bay Area, and Stanford University. NASA (or, more accurately, that part of the agency which Freeman Dyson called the "paper" NASA) provided funding to help support the study as did the American Society for Engineering Education. The original plan was to encourage faculty and students to learn "multidisciplinary engineering systems design" and take that knowledge back to their home institutions.[24] For Hopkins and Drexler, being limited to sketching and speculating had little appeal. "We wanted to actually do it," Hopkins recalled.

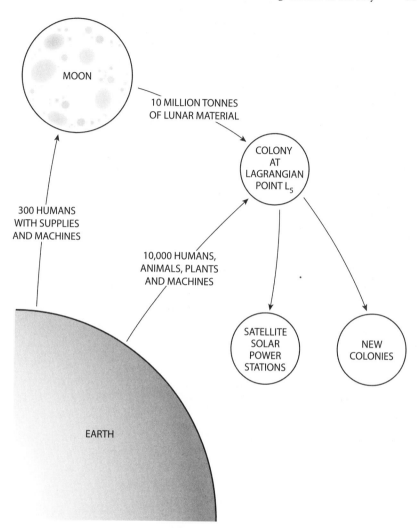

Figure 3.2 A schematic from a 1975 NASA study showing the "space colonization system." (Redrawn from an illustration courtesy of NASA.)

"The hell with the learning stuff. We wanted to get the best results we could."[25]

Now transformed from a pedagogical exercise into a full-blown ten-week design session, the study involved an interdisciplinary group of some few dozen engineers, physicists, social scientists, and students who started with O'Neill's basic concepts and improved them. Instead of a pair of rotating cylinders, for example, the group concluded that a torus or wheel-like structure about a mile in diameter would provide the most living space for the lowest cost. Spinning it about once a minute could generate an earth-like artificial gravity for its ten thousand inhabitants. Study participants also worked out detailed schedules, budgets, and milestone charts for accomplishing the entire system.

Although O'Neill was officially listed as the study's technical director, he didn't always agree with the participants' calculations or design decisions. For instance, a primary safety issue for living in space is shelter from cosmic radiation. Discarding O'Neill's belief that the settlement's structure would provide adequate passive protection, the Ames/Stanford group instead proposed a separate radiation shield whose mirrored surface would reflect sunlight to the habitat. After the study ended, O'Neill confided to a friend that it had been "a madhouse ... various evil characters tried to take the whole space-colonies concept away from me." It thus became "an awful struggle with all sorts of evil politicking in the background."[26] Richard Johnson, an engineer from Ames and one of the study's codirectors, remembered that the comprehensive design process was sometimes a "bitter pill" for O'Neill. "He kept on saying that it was not a dream," Johnson said. "He wanted it to happen in his lifetime."[27] Despite his objections, the study indubitably helped strengthen the technical soundness of O'Neill's initial design by forcing the continued rethinking of his ideas.

The study's participants paid close attention to how one could build sustainable closed-cycle ecological systems. With both the American and Soviet space agencies housing astronaut crews on space stations such as *Skylab* and the various Salyut platforms, this was a research topic scientists were keenly interested in. It also presented a far more complex problem, as small, closed systems are especially sensitive to environmental instabilities.[28]

Study participants were also interested in what *living* in a space settlement might be like (as opposed to just designing one) and considered the sorts of agriculture and animal husbandry that might make the most ecological sense. The curiosity some study participants had for communal living and libertarian-oriented governments blended with questions about the social and psychological effects of long-term space habitation. Would, for example, living in space be akin to *shimanagashi*—a form of banishment used during feudal Japan? One physics professor cited the presence of children and spontaneity as necessary for a settlement's success. "You must have part of your environment which is not under your control," he said. "It gives you a sense of reality which you need for mental health."[29]

When the Ames/Stanford summer study concluded, NASA released its initial findings, which roused more media interest. Aside from the technical details, reporters were intrigued by the idea that space colonization might be a "paying proposition" that would take advantage of the "scientific information returned to us by Apollo." It would also offer a "way out from the sense of closure and of limits which is now oppressive to many people on Earth." For a world that had "lost its frontiers," an expanded human presence in space could offer a "sense of hope and of new options and opportunities."[30] Speaking to the report's techno-enthusiasm, O'Neill told a local New Jersey paper: "It's as though this idea is filling a number of very deep needs. After Watergate and all the problems we are faced with, it seems to be a bright light, a cause for optimism."[31]

The 1975 workshop was the first of three such summer studies NASA helped sponsor. Each of them grew in size and produced even more detailed designs as O'Neill and his colleagues addressed specific engineering questions. In 1976, their work focused, for instance, on the mass driver's design and ways to chemically process the lunar materials it would hurl out to space. The following summer, participants gave detailed attention to ecological issues such as "regenerative life support systems and controlled environment agriculture."[32] These summer studies convinced O'Neill ever more strongly that the technical basis for his visioneering was sufficiently sound. The next area he needed to shore up, as a step toward building public interest, was the economic rationale for the humanization of space.

Let the Sunshine In

The financial questions raised by people such as Mark Hopkins prodded O'Neill to alter his strategy for the humanization of space. As O'Neill perceived a glimmer of interest from government and professional organizations, he opted for a more "conservative and pragmatic approach" in order to attract broader support.[33] One step was rebranding the Princeton conferences as meetings about "space manufacturing facilities" instead of space settlements. And, rather than advertising the events with futuristic images of space settlements, public announcements showed, for instance, the university's staid Woodrow Wilson School.

More importantly, O'Neill sought to enhance the immediate relevance of his ideas by pointing to tangible environmental and economic returns for people, politicians, and industries. As a result, starting around 1975, O'Neill included what became known as "satellite solar power stations" into his visioneering for space-based settlements and industry. In time, the idea of space-based solar power became a central pillar on which O'Neill rested his entire vision for the humanization of space.

Like many ideas that technology enthusiasts embraced, O'Neill's new tack had precedents in science fiction. For example, Isaac Asimov's 1941 story "Reason" features two astronauts stationed on an orbiting platform who are forced to negotiate with the intelligent, religiously inclined robot that controls the station's transmitted power beams. It wasn't until 1968, however, when *Science* published a research article exploring how vast amounts of solar power could be collected and transmitted to the earth without wires, that the idea got more widespread attention from the scientific community.

Peter E. Glaser, a Czech-born mechanical engineer, wrote the article that inspired O'Neill.[34] Glaser's idea itself had a long gestation period. It started when Glaser, who was working at a Boston-based management consulting firm, learned that William Brown, an electrical engineer at Raytheon, had invented an "Amplitron." This device could increase the power of microwave transmissions.

It quickly found military and civilian applications including helping improve the television images the Apollo missions broadcast from the moon. Brown was especially intrigued by the possibility of wirelessly transmitting electric power over great distances. A key part of this setup was a "rectenna," which absorbed microwave beams and converted them to direct current. In 1964, Brown even went on national television and demonstrated a small model helicopter powered by electricity transmitted wirelessly to it.[35]

Meanwhile, photovoltaic devices had improved dramatically in the 1960s because of the needs of the space program. But why put these solar cells in *space* instead of on the ground? For one reason, the earth's atmosphere absorbs a good deal of the sun's light. Therefore, a space-based collector could produce more power compared with one of the same size on the ground. Glaser took advantage of these facts and expanded on Brown's research to propose a system of orbiting geosynchronous satellites. At least one satellite would always be illuminated by the sun. These could collect solar energy and convert it to microwaves. Once the microwaves were beamed to earth, special antennae could collect this radiation and convert it back to electricity in order to supplement, perhaps even replace, traditional electric plants powered with fossil fuels. A visioneer cut from the same cloth as O'Neill, Glaser received a patent for his plan in 1973, regularly testified before congressional committees, and promoted his concept at conferences throughout the 1970s and '80s.[36]

Glaser's work convinced O'Neill that solar power beamed from space was technically possible. He wrote an article linking Glaser's ideas to his own and submitted it to *Science*, which responded swiftly and positively. "Space Colonies and Energy Supply to the Earth" appeared in late 1975.[37] The journal even gave O'Neill a coveted cover illustration that featured an artist's rendition of a space settlement's interior. Using the same tactic of coupling calculations with extrapolations, O'Neill described how people at a space manufacturing facility (he was already shying away from the word "colony") could build orbiting solar power stations. O'Neill argued that this approach had two advantages. Space-based power stations would be cheaper to build if the construction materials didn't have to be blasted up out of earth's gravity well. Even more importantly, solar power beamed back to earth would be a valuable commodity that would justify the economic investment in outer space.

For the rest of his life, O'Neill wove Glaser's concepts into his evolving grand vision for the humanization of space. It was exactly the type of big engineering project that appealed to the physicist. It provided O'Neill with a rationale for space development, and it meshed space exploration with environmental and societal needs. Current events also gave O'Neill a new social and economic frame in which to situate his vision. In the 1970s, industrial competition from Japan and Europe began to hammer the American economy. More critical was the oil shock that came in October 1973 after the Yom Kippur War. By the time the OPEC embargo ended in March 1974, the price of oil had quadrupled. Although oil prices eventually stabilized (before soaring again in 1979 after the Iranian Revolution), the crisis reminded politicians and consumers of the pessimistic scarcity scenarios in reports like *Limits*.[38]

But unlike some of the warnings *Limits* gave, which seemed abstract because they might not occur for decades, energy shortages produced real and immediate effects for American consumers, who experienced a national speed limit, lines at gas stations, rising home-heating prices, and even bans on Christmas light displays. A few weeks after Congress approved billions of dollars to build a trans-Alaska oil pipeline, President Nixon addressed the national energy crisis in a major speech. Invoking the Apollo program and the Manhattan Project, he challenged American scientists and engineers to join a new effort, christened Project Independence, to "meet America's energy needs from America's own energy resources."[39] Nixon eventually estimated that achieving this goal might require some $500 billion of federal and corporate investment over the next decade.[40] As O'Neill and Glaser pitched it, unlimited solar power beamed down from space was a bargain.

Journalists writing about O'Neill's ideas started to highlight the connection between future space settlements and obtaining a potentially limitless supply of clean energy. For example, when an article in *Time* reported on the 1975 space settlement study, it noted that a prototype satellite solar power station and accompanying space habitat could be built for a fraction of the cost of Project Independence.[41] NASA began providing O'Neill with a very modest amount of funding to explore his ideas. While this did not mean that space-based factories and settlements were a priority for either the space agency or the aerospace industry, NASA started to include concepts derived from O'Neill's work in some of its long-range planning exercises.[42] This, again, highlighted the distinction between the "paper" and "real" NASA that Dyson had already recognized and which O'Neill was now learning about.

O'Neill's education in politics continued when, in July 1975, he appeared before a congressional subcommittee for series of discussions titled "Future Space Programs." For five days, politicians heard speakers such as Carl Sagan and Isaac Asimov speculate on long-term plans for space exploration. Norman Cousins, editor of the big-ideas magazine *Saturday Review*, led off the hearings by explicitly countering the pessimism that permeated reports like *The Limits to Growth*. More than three years later, the Club of Rome's warning still provoked strong responses, as its dire simulations ignored the potential appearance of someone "capable of generating ideas that can lead to great change."[43] The future, Cousins claimed, "will be a product of the human mind, rather than of the computer."

O'Neill couldn't have agreed more. In his testimony, he hitched America's future energy needs to an expanded human presence in space. His colorful slides depicted artists' renditions of space settlements, mass drivers, and satellite solar power stations. He even brought a small home-built model of a space settlement. Just as Chesley Bonestell and other artists helped normalize the idea of space travel for people in the 1950s, these props helped show space habitats as something that could be comfortably familiar. One engineer recalled that this "was a far cry from the sterile presentation of a handful of astronauts circling earth in a metal marvel," exactly the goal O'Neill wanted to achieve as he promoted the humanization of space.[44]

Aided by the optimistic associations O'Neill made between space settlements and earthly needs along with scores of newspaper articles, magazine stories, and illustrations, the idea of space settlements was quickly spreading well beyond the original core group of future-minded students and professors. Ignited by his lectures and articles, public interest in O'Neill's ideas was rushing toward liftoff.

"L5 by '95!"

The pro-space movement that spun off from O'Neill's visioneering did not originate at college campuses, public marches, or cafeteria sit-ins. Instead, it sprang out of the cramped Tucson office of a small electronics business owned by H. Keith and Carolyn Henson.

Keith Henson was a self-described army brat with a long-standing interest in science and technology. Born in 1942, Henson attended the University of Arizona in the 1960s, where he did coursework in technical fields like electrical engineering. Henson had a talent for working with his hands—one writer, with some hyperbole, described him as "a guy who could take the world apart with a screwdriver."[45] He also cultivated a reputation for controversy (later in his life, Henson was jailed after a legal battle with the Church of Scientology). Carolyn Henson, née Meinel, had grown up with space and the stars. Her parents, Marjorie and Aden Meinel, were professional astronomers. After a peripatetic childhood, Carolyn moved to the Tucson area in the late 1950s when her father, an iconoclastic optical engineer, became the director of the Kitt Peak National Observatory. Carolyn, who also had an engineering background, was interested in women's liberation and environmentalism, and both she and Keith had opposed the Vietnam War.

The Hensons were an energetic and opinionated (some detractors also used words like aggressive and abrasive) couple who shared a pronounced enthusiasm for science fiction writers like Heinlein, Clarke, and Asimov. Like many other young American couples their age, the Hensons, whether for philosophical or financial reasons, strove to be more self-sufficient. Their rambling house, located just off the university campus, had a large garden and was home to an array of chickens, goats, and rabbits. The Hensons' house offered visitors unconventional entertainment options. A network of underground tunnels ran beneath it and, instead of a television, the Hensons built a Tesla coil that created entertaining displays of electrical sparks. Meanwhile, the open desert spaces around Tucson allowed them to indulge a fondness for recreational explosives by reenacting scenes from Tolkien's *Lord of the Rings* with homemade pyrotechnic devices.

The Hensons learned about Gerard O'Neill via his 1974 *Physics Today* article. Intrigued by his vision for the humanization of space, the Hensons contacted the physicist in Princeton and inquired about what future space settlers would eat. They drew O'Neill's attention to their own family's "integrated lifestyle" with its homegrown animals and backyard garden. In response, O'Neill invited them to the 1975 Princeton conference, where they gave a talk called "Closed Ecosystems of High Agricultural Yield." Their paper drew on work by University of Arizona researchers that suggested rabbits fed an alfalfa diet could be an ideal high-protein food, something Carolyn demonstrated at the 1975 Ames/Stanford summer study by serving deep-fried rabbit and homemade goat cheese.

Inspired by these meetings, the Hensons turned their attention to building a grassroots pro-space organization. Like many other young technology enthusiasts, they questioned the *Limits* thesis and found NASA's plans for the post-Apollo era uninspiring. Moreover, like O'Neill and Freeman Dyson, they believed that the humanization of space would offer places for social experimentation. Carolyn predicted that "some creative anachronism people" would want to participate and imagined parallels with the earlier formation of Amish and Quaker communities as space settlers fabricated "their own little pocket of history to live in."[46]

The 1970s were, of course, what Tom Wolfe castigated as the "Me Decade." This judgment, critics said, was reflected in some baby boomers' elitist and self-centered attitudes. Correspondingly, other motives spurred the Hensons as well.[47] "We realized that if we waited around for the jobs to open up, it would never happen," Carolyn Henson told an interviewer in 1978. "The only way we're going to get jobs in space is to do it ourselves!" Moreover, life might simply get "very BORING if we stick around on this planet too long." Her husband was even more direct in his assessment of what space exploration could offer adventurous self-seekers: "We want to GO! And become millionaires and come back to earth to visit!"[48]

The Hensons launched the L5 Society in August 1975. The name, naturally, came from the Lagrangian point O'Neill proposed as the location for the first space settlements. From the outset, L5's founders harbored ambitious plans—to "educate the public about the benefits of space communities and manufacturing facilities," to serve as a clearinghouse for information and news, and to help raise funds for further work, especially if government funding was unavailable. But L5 professed an even grander goal. At some point in the not-so-distant future, the Hensons imagined that L5ers would convene at an actual space settlement for one final meeting and then disband, their mission accomplished.[49]

L5's membership grew gradually. Dues for joining the group were reasonable: $10 for students, $20 for everyone else. O'Neill shared the mailing list for his own newsletter as a way of helping the space pioneers drum up interest. Meanwhile, talks by L5ers at science fiction conventions, combined with grassroots proselytizing, expanded the membership. By 1977 some 1,700 people had joined, and the number climbed to over 4,000 by the time Ronald Reagan became president in 1981. Eventually there were several dozen local and national groups oriented in some way toward educating the public on space exploration's importance, and surveys estimated this broader community of technology enthusiasts had over 36,000 members.[50]

To help spread the group's message, the society encouraged members to make photocopies of the L5's monthly newsletter and share them with friends. Over time, Carolyn Henson transformed *L5 News* into a sophisticated vehicle, its slickness even more remarkable given the group's limited budget and publishing tools. Full of illustrations, some in color, *L5 News* relentlessly argued

Figure 3.3 Cover from one of the L5 Society's early newsletters, showing an artist's conception for a spherically shaped space colony. (Image by Rick Guidice, courtesy of NASA, NASA ID number AC76-0965.)

that an expanded human presence in space, with ordinary citizens leading the way, was vital for the future. *L5 News* typified an emerging underground and niche-oriented West Coast "print culture."[51] Many of these publications promoted particular counterculture lifestyles and activities—meditation, vegetarianism, handmade houses—to a relatively small but dedicated readership. L5's newsletters suggest some hybridized version of this, as they showed an alternative and ostensibly self-sufficient lifestyle of the future yet one mediated through some of the most complex technologies in existence.

Local chapters of L5 sprang up across the United States and overseas. West Germany, Finland, and Canada all had groups at some point. O'Neill's vision proved especially intriguing to young adults, and L5's chapters were, not surprisingly, often associated with universities and colleges. The University of Texas in Austin, for instance, hosted an active chapter that met regularly. Harlan Smith, an astronomy professor with a passion for popularizing science, was a member and later served on L5's board of directors. Another member, Deborah Byrd, soon started writing and producing *StarDate*, an Austin-based telephone message service that, in 1978, became a nationally syndicated radio segment on space and astronomy topics.

Interest among young people wasn't limited to college students. At John Muir High School in Pasadena, Taylor Dark III and a few friends established their own L5 chapter. The group's thirty members included Carl Feynman, son of Nobel laureate Richard Feynman, and it took advantage of its proximity to Caltech and the Jet Propulsion Laboratory to recruit speakers.[52] One of the mimeographed newsletters that Dark and his friends put together noted that their chapter was planning to visit a local aerospace factory to see a space shuttle prototype and that it would also participate in an upcoming fair for solar power and other alternative energy sources. Space hardware, environmental concerns, political activism—all were part of the teens' technology-related interests.[53]

L5's orientation toward grassroots activism proved attractive to many members as well. James C. Bennett, for example, recalled hearing O'Neill speak at the University of Michigan, where he was a student. The physicist's optimistic message prompted him to join L5. Bennett, inspired by Heinlein as well as New Right libertarianism, wanted to "extend the narrative of settlement and expansion of America into space."[54] As a science fiction fan, Bennett didn't see the idea of space settlements as novel in its own right. But many sci-fi readers were, Bennett recalled, "notoriously uninterested in doing anything." L5, in contrast, "brought out the people who wanted to do something."[55]

It is difficult to construct an exact and detailed demographic picture of the L5 Society's membership. But evidence shows that it was especially strong in California, Washington, Texas, and Arizona, states that were part of the postwar "Gunbelt," where the nation's defense contractors were concentrated. As one would expect, there were more extensive pockets of support near universities and centers of aerospace activity. Members, according to L5, tended to be "young, well-educated, and receptive to new ideas."[56] More than two-thirds were under the age of thirty-five, most were single, and about 70 percent had college degrees. O'Neill's hope, however, that the humanization of space would attract a broader swath of humanity remained largely unfulfilled. Although women such as Carolyn Henson took on some prominent roles—under her editorial guidance, "piloted spaceflight" as opposed to "manned spaceflight" became a preferable term in L5's newsletters—the group's members were overwhelmingly well-educated and middle-class white men.[57] Americans' "love affair" with technology had long been interpreted as a male domain, and enthusiasm for space colonies proved no exception.[58]

Despite the Hensons' bravado, space colonies were not making them rich, and L5 continued to operate on a shoestring budget. Several times, "angel donors" stepped forward at key points and gave large contributions to keep the newsletter in print or to send an L5 member to an important conference or meeting. Barbara Marx Hubbard, an early L5 board member, donated generously as did William B. O'Boyle, an independently wealthy and conservative investor from New York City. The humanization of space enthralled O'Boyle such that, in one six-month period, he gave L5 close to $10,000 (equivalent to some $40,000 in 2012).[59]

In the pre-Internet era, the L5 Society spread its message using more old-fashioned means. Besides the newsletter, the society organized a phone network that could alert members quickly about important issues. T-shirts and buttons broadcast the L5 message, and the Hensons invited members to take part in a bumper-sticker contest. "L5 by '95" spoke to the group's long-term goal, while "Space Power—L5" connected the group with satellite-based solar power plans. Entries like "Sex at Zero-G" and "L5—Let's Get High" reflected interests more connected with youthful rebellion.

Of course, one might not be able to actually *build* a space colony, but one could surely *write* about what living in it would be like. L5's Tucson office served as a clearinghouse for a wide array of information about the technological future. Members could, for instance, buy books by authors such as Hubbard, Asimov, or Timothy Leary. L5 also offered reprints of articles and posters and 35-millimeter slides that members could purchase to use in public presentations and recruitment talks. L5's newsletter also promoted books on space settlements and manufacturing by people like O'Neill, Thomas Heppenheimer, and G. Harry Stine. In the process, L5 Society helped establish a canon—an appropriate term given the near-religious fervor of some pro-space activists—of books, articles, and other publications as well as a community of people who shared its technological enthusiasm.

With ambitions for catalyzing a pro-technology social movement, the L5 Society promoted a broad set of issues connected, directly or indirectly, with space exploration. Consider lyrics written by William Higgins and Barry Gehm, two science fiction fans who met at Michigan State University while in graduate school. Higgins introduced their song at a Miami Beach science fiction convention in 1977 and, over the years, the two men performed it in a variety of venues including "a wading pool filled with dry ice, illuminated by laser beams." Composed in the vein of 1970s-era "filk music"— folk music that has a sci-fi theme—and accompanied by Higgins' ukulele, the song was meant to be sung to the tune of "Home on the Range":

> Oh, give me a locus where the gravitons focus
> Where the three-body problem is solved,
> Where the microwaves play down at three degrees K,
> and the cold virus never evolved.
>
> CHORUS: *Home, home on LaGrange,*
> *Where the space debris always collects,*

We possess, so it seems, two of Man's greatest dreams:
Solar power and zero-gee sex.

We eat algae pie, our vacuum is high,
Our ball bearings are perfectly round.
Our horizon is curved, our warheads are MIRVed,
And a kilogram weighs half a pound.

> *(chorus)*

You don't need no oil, nor a tokamak coil
Solar stations provide Earth with juice.
Power beams are sublime, so nobody will mind
If we cook an occasional goose.

> *(chorus)*

I've been feeling quite blue since the crystals I grew
Became too big to fit through the door.
But from slices I sold, Hewlett-Packard I'm told,
Made a chip that was seven foot four.

> *(chorus)*

If we run out of space for our burgeoning race,
No more Lebensraum left for the Mensch,
When we're ready to start we can take Mars apart
If we just find a big enough wrench.

> *(chorus)*

I'm sick of this place, it's just McDonalds's in space,
And living up here is a bore.
Tell the shiggies "Don't Cry," they can kiss me goodbye,
'Cause I'm moving next week to L4.[60]

The lyrics captured all the eclectic possibilities that attracted people to O'Neill's visioneering: alternative energy, the idea of making money by manufacturing goods in space, the ideal of raising one's food (even if it was algae pie), and even the possibility of some erotic experimentation. Even though Higgins and Gehm were gently poking fun at the enthusiastic ambitions of space colony enthusiasts, L5 members embraced the song as their "national anthem" and, over the years, the lyrics circulated widely throughout the science fiction and space enthusiast communities.[61]

Still, scenarios of a restricted future remained both a durable shibboleth and a stimulant for debate that continued to motivate many L5 members. When a Texas oil millionaire organized a conference on the question of limits, L5 sent some members to participate. There they mingled with prominent economists and pro-technology futurologists such as Herman Kahn, famed for his writings on nuclear war.[62] One person who attended, L5 member and physicist J. Peter Vajk, went on to write his own riposte to *Limits*. Called *Doomsday Has Been Cancelled*, Vajk's book blended his own philosophical views—the "conscious and deliberate creation of the future"—with a pro-space and anti-*Limits* argument.[63]

Given members' enthusiasm for Glaser's satellite solar power stations and the prominent media coverage that anything energy-related received in the mid-1970s, the *L5 News* regularly featured stories about American energy policy. Carolyn Henson's astronomer parents helped stimulate their daughter's enthusiasm for space-based power stations and massive engineering projects in general. In the 1970s, they advocated for the construction of vast solar farms in the desert Southwest that would produce electricity on a grand scale.[64] Although the Meinels supported solar power, the plan they favored bore little resemblance to the small-scale "appropriate technologies" championed by most traditional environmentalists. Space colonies and giant solar power stations represented an idiosyncratic solution to problems that some Americans were already tackling with backyard windmills and rooftop solar panels. As one L5 supporter from New Jersey wrote, "energy and economics can stand alone as justification for the colonization of space." Sentiments such as this reveal that many pro-space enthusiasts saw nothing schizophrenic in pursuing environmentally friendly goals via resource-intensive technological solutions.[65]

In fact, a strong environmentalist theme ran through L5's early publications. It insisted that concern for the planet and "big technology" could coexist and complement each other. For example, L5 filled empty spaces in its newsletters with sayings like "Declare the earth a wilderness area" and encouraged members to buy bumper stickers with a picture of Spaceship Earth and the simple slogan:

If you love it . . . leave it.
L5.

The first few years of the newsletter also featured regular articles by Norie Huddle, a young Peace Corps veteran who had returned from South America to become an activist interested in environmentalism and alternative futures. Besides serving on L5's board, Huddle traveled to Vancouver in May 1976 to attend the United Nations' Habitat Forum. There, she met with activists concerned about urbanization's deleterious effects on ecosystems and distributed an L5 pamphlet called *Human Settlements in Outer Space* that was printed in English, French, and German.[66]

L5 attracted people from across the political spectrum. In its first few years, the group received endorsements from presidential candidates Morris Udall and Lyndon LaRouche, conservative Arizona senator Barry Goldwater, libertarian sci-fi writer Robert Heinlein, Norie Huddle (L5's "token environmentalist," the group claimed), and New Age futurist Barbara Marx Hubbard. Although celebrities like Heinlein and Goldwater didn't play a major role in running the L5 Society, they helped promote the group's agenda through promotional pieces and endorsements. As a result, the citizens' pro-space movement presaged the odd political alliances that emerged two decades later when left- and right-wing writers and political leaders united in their enthusiasm for the Internet and the opening of the new "electronic frontier."[67]

Keep on SMI²LEing

O'Neill expressed mixed feelings about the tide of interest he stimulated. On one hand, he recognized the importance of grassroots activism. His talks and articles encouraged citizens to get involved, and he called the appendix of his 1977 book *The High Frontier* "Taking It to the People." At the same time, O'Neill was wary about the messianic fervor some L5ers expressed. Over time, the Princeton physicist's caution turned to deliberate avoidance, and eventually he referred to L5 simply as "those people in Arizona."[68]

Several factors drove O'Neill's gradual estrangement from L5 and the other grassroots groups his work inspired. In 1976, O'Neill began collaborating with Henry Kolm, the MIT physicist whose work on electromagnetically levitated transportation had inspired the idea for the mass driver. Their partnership coincided with O'Neill receiving a prestigious visiting professorship at MIT. Funding from NASA for work on the mass driver started at about $25,000 annually and rose to some $250,000 by 1980.[69] This was enough to support Eric Drexler and several other students from Princeton and MIT as they built a prototype. This test device could accelerate a payload to eighty miles an hour in a tenth of a second, so fast that when the public television science program *Nova* showed the mass driver's public debut, the small payload seemed to vanish and then reappear a split second later some fifty feet away.[70] Eager to capitalize on their success, O'Neill and Kolm sought more funding to build an even bigger version ,which necessitated that their project appear as "scientific" as possible.

Another factor that gradually separated O'Neill from L5 stemmed from his personal career experiences. O'Neill's exodus from conventional physics research certainly raised eyebrows throughout Princeton's Physics Department, especially after the soft-core pornography magazine *Penthouse* published an interview with him.[71] The physics community had adopted O'Neill's storage ring concept only after he spent years doing grueling experiments and promotion, and he never believed he received the credit he deserved. Now, with his new idea making headlines, O'Neill was eager to maintain control. But, as his visioneering gained international publicity and attracted devotees, this became increasingly difficult.

Maintaining intellectual integrity was especially important to O'Neill. He remained a tenured professor in one of the world's best physics departments, and journalists' accounts that misquoted him or sensationalist articles that bungled technical details dismayed him. In his newsletters, sent out semiregularly to hundreds of interested colleagues and citizens, he highlighted articles that deserved kudos for their accuracy.[72] Meanwhile, his personal files were filled with articles about him that he clipped and annotated with corrections.

Compared with the typical L5 member, O'Neill, with his Ivy League affiliation and personal ties to Nobel laureates and NASA officials, remained part of the mainstream "science Establishment," even if his ideas did not always attract people with similar credentials. The diverse demographics and potential mass appeal of the L5 Society concerned him as the group attracted members from society's fringes. Letters from "kooks and con artists" appeared regularly at L5's office, and one deluded enthusiast even showed up in Tucson wanting to turn his pool table into a starship.[73]

If O'Neill was discomfited by the isolated oddballs who represented the fringe of L5's membership, he was even more wary when Timothy Leary, once labeled by Richard Nixon as the "most dangerous man in America," embraced space colonization. In April 1976, after his release from prison, the former Harvard professor turned LSD guru began promoting another trip: humanity's exodus into space. This was part of a larger vision for "Space Migration, Intelligence Increase, Life Extension," which Leary cheerily branded SMI²LE.

Leary first began thinking about space migration in the early 1970s while he languished in California's penal system on drug charges. The arrival in 1973 of comet Kohoutek, which some hippie cultists saw as a harbinger of doom, inspired Leary. He wrote a tract titled *Starseed*, which he then "transmitted" from the "black hole" of Folsom Prison.[74] In it, Leary claimed that he was preparing "a complete systematic philosophy: cosmology, politic, epistemology, ethic, aesthetic, ontology, and the most hopeful eschatology ever specified."[75] Kohoutek (the "starseed" of Leary's title) was proof that a "higher Intelligence has already established itself on earth, writ its testament within our cells, decipherable by our nervous system. That it's about time to mutate. Create and transmit the new philosophy. . . . Starseed will turn-on the new network."

After he read Leary's short opus, Robert Anton Wilson, a Bay Area science fiction writer, decided to organize a small group. Taking a cue from Leary's missive, he called it The Network. Known for his *Illuminatus* books, which mixed occult themes, conspiracy theories, and wildly discontinuous narratives, Wilson became an aficionado of SMI²LE. The Network's logo even featured the words "neurologic-immortality-star flight" around a stylized yin-yang symbol.[76] Wilson had an especial reason to be drawn to Leary's techno-evolutionary optimism. After the murder of his teenage daughter in a robbery, Wilson had her brain cryogenically preserved—frozen—in the hopes that one day she could be cloned.[77] Leary's SMI²LE offered the bereaved writer some measure of hope.

Just before his release from prison, Leary claimed to be spending "twelve hours a day reading and writing. Lots of science."[78] Through Stewart Brand's new magazine, *CoEvolution Quarterly*, Leary learned about O'Neill's "great space colony revelation." He concluded that "only in space can we take the next steps in our evolution."[79] Whether he truly believed this or was just riding the coattails of O'Neill's publicity is hard to tell. But, in any case, Leary began heaping superlatives on O'Neill, calling the physicist, for example, the "most important human being alive today."[80]

After getting out of jail, Leary, now branding himself a "hope fiend," started promoting both O'Neill and his own new ideas. In the summer of 1976, The Network held its first "Starseed Seminar."[81] Featured guests included Leary, *Doomsday Has Been Cancelled* author Peter Vajk, and advocates from the Bay Area Cryonics Society. Another person pulled into Leary's orbit was George Koopman, a former Vietnam-era intelligence analyst. In the mid-1970s he embraced the growing New Age movement, befriended Leary, and joined L5's board of directors.[82] Koopman, who had both a long-standing interest in space and family money, was a perfect addition to Leary's coterie.

Several young underemployed physicists rounded out the guest list for SMI²LE-related events. Intrigued by the confluence of Eastern mysticism and quantum theories popularized in books such as Fritjof Capra's best-selling *Tao of Physics*, they gave their own riff on Leary's radical technological enthusiasm.[83] One of them, Jack Sarfatti, had ties to Werner Erhard, the millionaire founder of Erhard Seminars Training (i.e., "*est*"). For a healthy fee, *est* offered people a chance for self-awareness and personal growth. Sarfatti invited Leary to talk about SMI²LE at workshops at the Esalen Institute (once called "a Cape Canaveral of inner space") amidst Big Sur's rugged beauty.[84]

In 1977, a new book by Leary, with contributions from Robert Anton Wilson and George Koopman, gave an unruly articulation of Leary's ever-evolving message. In *Neuropolitics: The Sociobiology of Human Metamorphosis*, Leary said SMI²LE would enable humans to improve themselves via "neurogenetic evolution preprogrammed by DNA" just as "scientific and technological advances were making space migration a practical alternative to our polluted and overcrowded planet."[85] Overall, it blended philosophical antilimits musings, fiction, and political satire with whole sections wreathed in scientific jargon.

Leary believed the humanization of space à la O'Neill could provide all sorts of opportunities so long as it didn't become an "insectoid bureaucracy, or another Alaskan pipeline controlled by the oil politicians."[86] Space itself could be the new petri dish for social and biological experimentation. "Housing not monastic astronauts but families, tribes, human communities," space colonies would represent the "greatest evolutionary mutation since the ascent from the ocean to land."[87] Readers of *Neuropolitics* also learned, among other things, why "dopers seem to prefer sci-fi" and how space colonies would permit sexual experimentation.[88] It's difficult to determine how people responded to Leary's books, but copies that circulated within the University of California's library system are heavily marked with phrases like "how to see the world," "TOWER OF BABEL," and "technological mysticism" scattered throughout.

From his home, tucked away in one of Los Angeles' steep, shaded canyons "where the migrants and the mutants, and the future people come from, the end point of terrestrial migration," Leary helped recruit people to the pro-space movement.[89] Taylor Dark recalled that, as a high school student, he attended one of Leary's SMI²LE talks, where he saw Timothy Leary bounding onstage singing the lyrics to the Bee Gees' hit song "Stayin' Alive."

Afterward, Leary offered Dark advice on how to promote space migration to friends and political leaders. Leary also became friends with the Hensons, who helped promote his books and ideas in L5 newsletters. For their part, L5 members were divided about taking Leary on board. Although he could generate publicity, some L5ers feared that, in Leary's hands, space migration might just become "a new way . . . to Turn On and Drop Out."[90]

No visioneer, Leary did no design or technical work to bolster his ideas, and O'Neill avoided direct associations with him. Nonetheless, Leary dropped the Princeton physicist's name into almost every exposition of SMI²LE and claimed "sexy Gerard O'Neill" was proof that the days of the "retiring, square, fuddy-duddy scientist" were over.[91] Leary's enthusiasm for "high orbital living" highlighted difficulties that visioneers have in controlling their message and ideas. It also showed how far O'Neill's humanization of space had migrated from its origins in Ivy League classrooms and NASA workshops. As the discussion continued to diffuse from lecture halls and sci-fi conventions to the coffee shops and hot tubs of coastal California, it mutated into debates O'Neill and his initial band of acolytes would not have imagined.

"Apocalypse Juggernaut Goodbye"?

In June 1975, Stewart Brand attended the World Future Society's general assembly in Washington, DC. The mood was dour. "Futurists were more interested in problems than solutions that year," he recalled. Few people showed up to hear Gerard O'Neill's more optimistic talk. Brand was one of them, and O'Neill's vision converted him "from mild interest to obsession."[92] That obsession would play out over the next few years as Brand catalyzed a debate among counterculture icons, environmentalists, scientist-celebrities, and technology enthusiasts.

Brand had been interested in space exploration since childhood, thanks, in part, to his mother's enthusiasm for the topic. After reading Tom Wolfe's essays in *Rolling Stone* (which presaged his 1979 best seller *The Right Stuff*), Brand came to see the astronaut as a seeker who blended the "unfashionable aesthetic" of astounding technological prowess with a personal voyage of discovery. "We were wrong," one of his diary entries said, "in perceiving the astronauts as crew cut robots."[93] His growing friendship with astronaut Russell "Rusty" Schweickart fostered this change in perspective. As a member of the Apollo 9 mission, Schweickart had spent more than two hundred hours in space. Schweickart was vocal about his "sensitivity to left-of-center issues" and spoke openly and emotionally about his spiritual experiences in space.[94] If Schweickart and others were right in thinking that space was "the new ocean," he asked himself "Where did that leave me?"[95] Although no visioneer, Brand possessed tremendous talents for organizing and networking, and his promotion helped O'Neill find a wider audience.

In 1974, after recovering from a near–nervous collapse brought on by the pressure and publicity that *Whole Earth* generated, Brand started to think more about space. He also launched a new publishing venture. An eclectic publication, *CoEvolution Quarterly* blended politics and a "western libertarian sensibility" with enthusiasm for alternative or "soft" technologies.[96] *CoEvolution*'s dominant theme was that some reconciliation of consumerism, environmentalism, and technology was possible. Its outlook was optimistic. As Brand said, chiding adherents to the *Limits* thesis, people who "organize their behavior around the apocalypse" would be lost when doomsday failed to happen.[97]

CoEvolution made its first big splash in the summer of 1975 when it presented Lynn Margulis and James Lovelock's "Gaia hypothesis." A controversial scientific expansion of the Spaceship Earth idea, it depicted the earth's biosphere as a complex and naturally self-regulating system. The same principles were at the heart of the rudimentary artificial environments O'Neill envisioned, and Brand in fact had extensive correspondence with biologists as to whether outer space might be a viable place to test Gaia-related ideas.

The success of the Gaia article prompted Brand to imagine *CoEvolution* maturing into a countercultural journal that focused on science and technology. Concerned that researchers with unconventional ideas had no venue, which resulted in "the shutting off of science from the rest of human discourse," Brand saw *CoEvolution* as a West Coast alternative to traditional Establishment science journals. Seeking new perspectives, he even contacted MIT's Philip Morrison, *Scientific American*'s book critic (and, as we saw, one of Eric Drexler's advisers), to pitch *CoEvolution* as an alternative for "good stuff you [*Scientific American*] can't use." Brand imagined his magazine eventually moving from "secondary to primary science material, in context with art, poetry, sex, and how-to."[98] This broad-minded approach sometimes created odd confluences. One *CoEvolution* reader, for instance, wrote Brand about Project Grendel, which had the goal of "studying and obtaining legal protection for the North American sasquatch."[99]

Scientific pretensions aside, *CoEvolution* maintained a focus on the same alternative and appropriate technologies that Brand championed previously with the *Whole Earth Catalog*.[100] Articles about simple, cheap, eco-friendly, and safe technologies—solar energy, recycling, composting, communal living—were especially popular. At the same time, *CoEvolution*'s readership included many science fiction fans, and Brand sought speculative contributions from futurists. O'Neill's vision for space settlements fit right in with *CoEvolution*'s editorial rubric. There was a personal affinity as well. Brand, contrary to expectations held by some in the waning counterculture movement, had (like O'Neill) served in the military. He also avoided traditional liberal politics in favor of a more nuanced libertarian philosophy. Brand had few sacred cows. His 1971 polemic called "Commune Lie" paraphrased Robert Heinlein's famous adage "there ain't no such thing as a free lunch" as it derided slothful escapists who took advantage of the era's communal living movement and gave nothing back.[101]

If Brand and other "cosmic cowboys" envisioned the American West as a tabula rasa that they could remake with windmills and hand-built homes, outer space offered even more possibilities but with "no conquering, no exploitation, no waste this time." *Whole Earth*, and now *CoEvolution*, were about communal living and shelter building as practiced by self-motivated people who made their own habitations. What Brand called "Free Space ... too big and dilute for national control" was the logical (if impractical) extension of these ambitions. For people interested in Gaia and earth-based systems ecology, experiments in which space settlements were treated as closed biological systems might help explore these ideas. The debate Brand fostered over space colonies was also an opportunity to test his famous assertion—"We are as gods and might as well get good at it"—about people's relation to technology. If nothing else, a dialogue about O'Neill's ideas could provide a "fresh angle on old problems" such as population growth, resource depletion, the energy crisis, and the sorts of technology that were indeed "appropriate."[102]

Like O'Neill, Brand believed that spending some tens of billions of dollars for a space settlement prototype might be worthwhile given what the United States was preparing to invest in energy independence. This was especially the case if O'Neill's ideas offered any escape from the predictions of *Limits* (a hopeful possibility Brand phrased as "apocalypse juggernaut goodbye.")[103] Brand also recognized the controversial nature of O'Neill's grand vision: "The man-made idyll is too man-made, too idyllic or too ecologically unlikely—say the ired. It's a general representation of the natural scale of life attainable in a large rotating environment—say the inspired. Either way, it makes people jump."[104]

Brand's admitted obsession filled the mail bin at his Sausalito office and the pages of *CoEvolution Quarterly* for more than two years. Starting with its 1974 "Winter Solstice" issue, *CoEvolution*'s articles and letters both attacked and praised O'Neill's ideas. *CoEvolution* featured space exploration on its covers twice, including one lovely view of a space settlement that looked strikingly like a miniature San Francisco. Few, if any, other topics received such prominent coverage in *CoEvolution* during this time. Brand himself tried to remain neutral, but his diary notes reveal his real loyalty. "Technology, kiddo," he wrote in his journal after seeing the first space shuttle. "This is to today what the great sailing ships were to their day." Making a dig at regressive commune dwellers, he admonished "Get with the program or stick to your spinning wheel."[105]

In response to the issue of space colonies, about four out of five correspondents viewed the humanization of space favorably. Most of the people who wanted to participate were college students; women were as much in favor as men; and the "most universally favorable group" was artists. Some *CoEvolution* readers imagined living in space as an extension of a "back to land" lifestyle that eschewed crowded and polluted urban locales for rural communes, social experimentation, and wilderness preservation. This was oddly ironic given that space settlements would be hugely intensive in terms of resource consumption and capital investment, and would require Apollo-like management to succeed. A few readers, seeing outer space as a last refuge, expressed fatalism, sadness, and even a sense of desperation: "Whatever I can do," said one, "may help my beautiful daughter to slip away from this failing civilization here on Earth."[106]

But what about the minority of correspondents who expressed disfavor? The grand vision of space colonies provoked outrage among these readers and, moreover, ignited their fears of technocracy. Spending such huge amounts of money to circumvent the planet's limits struck one reader as "well thought out, rational, very alluring," and also "quite mad." It appeared to follow the notion of a technological fix to its logical extreme. For these people, O'Neill's ideas violated E. F. Schumacher's "small is beautiful" philosophy and the ideals of small-scale appropriate technology they admired. Others detected signs of a massive new federal program and the military-industrial complex at work ("the same old technological whiz-bang and dreary imperialism").[107] A two-page illustration *CoEvolution* published captured readers' split opinions. One page showed an artist's colorful rendition of a spherical space settlement. The facing page presented a nineteenth-century photograph of a Native American couple who appeared to be gazing at it—the text added above the man's head said, "Goodbye. Good luck." The woman's reaction? "Good riddance."

But Brand ultimately wanted to spark a conversation that transcended the particular merits of space colonies in order to address more general beliefs and ideals people had about reconciling nature and technology. To achieve this goal, he solicited opinions from a veritable "who's who" list of 1970s-era environmentalists, both writers and activists.

Compared with the opinions of *CoEvolution*'s general readers, the verdict from the environmental movement's elite was resoundingly negative. Most scientists, advocates of "appropriate technologies," and other intellectuals Brand approached (some were his personal friends) voiced not only mistrust but also downright loathing of O'Neill's vision. E. F. Schumacher, for example, sarcastically offered to help fund the migration of every person he saw as standing in the way of solving the earth's environmental problems. Others offered less direct denunciations. Richard Brautigan, author of *Trout Fishing in America*, responded in verse with a message of caution:

> I like this planet
> It's my home and I think it needs our attention and our love
> Let the stars wait a little while longer
> They are good at it.
> We'll join them soon enough.
> We'll be there.[108]

Such statements, written for publication, often arrived in Brand's mailbox with more frankly stated private musings. Author and social critic John Holt peppered Brand almost obsessively with denunciations of O'Neill. He also chastised Brand's seeming abandonment of appropriate technologies. "Be at least as tough on space colonies as you would [when recommending] a Skil-Saw," he admonished. "After all, they stand to cost a lot more and hurt a lot more people."[109] Wendell Berry, a poet and environmentalist who wrote about small-scale agrarian lifestyles, told Brand: "I do not hate technology. I hate the misuse of technology. . . . You ask 'Who are we depriving by hijacking an asteroid?' I answer that 'we' are depriving me. . . . I want to know who issued Gerard O'Neill his strip mine permit."[110] To Berry, O'Neill was a salesman, uttering "every shibboleth of the cult of progress" while blinded by a naïve understanding of the American frontier story.[111]

The critics directed equally pointed barbs at Brand. One former *Whole Earth* editor called on him to not put the "gardeners at war with the space nuts" and to once again advocate "tools, not political abstractions."[112] Perhaps most cutting was this verse from one *CoEvolution* reader:

> It seems to be no secret
> Where all the flowers have went
> Yesterday's counterculture
> Is today's establishment.[113]

The intellectuals and activists who wrote Brand reflected persistent and widespread fears and pessimism about massive technologies such as nuclear power and the Apollo program. Brand admitted he was trying to provoke colleagues who had "become too predictable of late, too smug, certain ... and unimaginative. We have come to love our famous problems (population, inequity, technology, etc.) and would feel meaningless if they went away. That's a lousy design posture."[114] Brand's provocation succeeded. His "space colonies debate" highlighted the tensions between people who wanted to resolve the dilemmas *Limits* predicted through political change and small-scale technologies versus those who wanted the future that O'Neill's visioneering promised.

Achieving Apogee

Whether one supported or opposed visioneering for the humanization of space, 1977 marked its apex. That year saw the publication of O'Neill's book *The High Frontier: Human Colonies in Space*. Translated into at least five languages, its blend of science popularization, engineering possibilities, and fictional accounts of life at a space colony made it a best seller and brought O'Neill more media attention. Reviews of the book were generally positive. Even critics who found O'Neill's enthusiasm naïve praised his ability to speak to both professional colleagues and the general public.[115] Phi Beta Kappa gave the book a strong endorsement with its 1977 Award in Science, presented to people who made "outstanding contributions" to the "literature of science." Previous winners included Barry Commoner and chemist Linus Pauling.

The High Frontier presented the holistic vision that O'Neill had been developing since 1969. The book became part of the growing canon of works read by technology enthusiasts in the late 1970s. It was also used as a textbook for some university courses that transformed O'Neill's ideas into pedagogical tools. One school in Florida, for example, offered a new course called The Colonization of Space as a way of introducing undergraduates to "current topics in science and technology." This attempt to jump-start a sometimes unappealing curriculum reflected a more general trend of drawing students to science through nontraditional approaches such as the "Zen of physics."[116] Professors' interest in the topic also coincided with the growth of "science, technology, and society" programs at U.S. and European schools.

Soon after *The High Frontier* appeared, O'Neill announced that he had established the Space Studies Institute. Based in Princeton, the small nonprofit organization began to sponsor research on critical bootstrapping technologies for space exploration "no matter what political winds might blow through Washington."[117] Foremost among these were the mass driver, which was in need of continued development, as well as techniques for processing lunar materials. By the end of that year, O'Neill's institute had raised some $100,000 from private sources—close to $400,000 in today's terms—and, by the decade's end, it claimed a mailing list of over a thousand potential donors interested in space settlements.[118] O'Neill conceived of SSI partly as a way to avoid having to rely on NASA. His relationship with the space agency had been fraught from the beginning. Although it had modestly supported his mass-driver work, he found the agency's administration cumbersome and its long-term vision for human spaceflight lacking. The Space Studies Institute offered O'Neill a chance to put his money where his mouth was by trying to develop a truly citizen-based space initiative.

Soon after O'Neill's book appeared, millions more people learned about him when CBS's news show *60 Minutes* did a segment on the humanization of space. It immediately drew its viewers in by segueing between space battle scenes from *Star Wars*, which was still in theaters, and clips of the space shuttle *Enter-*

prise's test flight. Reporter Dan Rather described space colonies as a solution for "people looking to get away from an overcrowded earth" and perhaps a way to recapture elements of "small town America." The piece showcased O'Neill, who described himself as a "romantic, an idealist, and a practical physicist and engineer." A succinct definition of a visioneer, in other words. So far as his bold plans, O'Neill told Rather that he hoped that "whenever the romance starts getting out of hand, I hope the physics and the engineering pulls it back to reality."

The *60 Minutes* program also highlighted the diverse communities interested in or outraged by the idea of building "man-made planets." Carefully worded comments from NASA representatives were interspersed with disparagement from critics like John Holt. Meanwhile, L5ers filmed at a Tucson science fiction convention enthused about the possibility of building a place in space for Americans of all political, sexual, and lifestyle orientations. Looking further out on the political spectrum, *60 Minutes* reported that Black Panther leader Eldridge Cleaver was a supporter as was Timothy Leary, who was shown pitching SMI²LE to a group of young adults. "How many of you," the effervescent Leary asked, "would like to live in space and live forever?" What did O'Neill think of the diverse supporters he attracted, reporter Rather inquired? The physicist demurred, saying that the best path would be if space settlements were an idea "forced on the government . . . by the people and not the other way around."[119]

The CBS show also revealed that NASA was funding O'Neill. This news, coupled with pictures of NASA-built models, implied the agency was seriously considering O'Neill's ideas. Although the actual amount NASA gave O'Neill was minuscule compared with the agency's multibillion-dollar overall budget, the news incensed Wisconsin senator William Proxmire. Famed for his Golden Fleece awards given to protest government waste, he became known as "Darth Proxmire" to L5 supporters. Proxmire promised "not a penny for this nutty fantasy" and suggested instead that an irresponsible NASA should have its budget cut. Not surprisingly, the space agency backpedaled from announcing any new "high-challenge, highly-visible space engineering" initiatives.[120]

Proxmire's condemnation aside, some politicians noticed the public interest O'Neill's ideas generated. California's governor Jerry Brown was also mindful that aerospace companies, and technology-oriented firms in general, were a mainstay of his state's economy. Already informally advised by Stewart Brand, Brown asked Rusty Schweickart to serve as his science adviser. Encouraged by the two men, Brown started advocating a "California Space Program." One feature of his initiative would be a dedicated satellite to improve communications within the Golden State and facilitate environmental monitoring.[121] This plan was later lampooned by Chicago columnist Mike Royko, who termed Brown "Governor Moonbeam." Nonetheless, Brown began to see space exploration as socially useful and economically relevant, ideas that O'Neill's vision and celebrity helped reinforce.

While attending a space shuttle demonstration, Jerry Brown noted that space "brings people together . . . we had [a] Birch Society Congressman . . . Barry Goldwater, Tim Leary, Jacques Cousteau."[122] But, as members of the L5 Society experienced firsthand, an expanded human presence in space meant different things to different groups. Such a diverse coalition could not last for long. The next year, when the Golden State held its second Space Day, it was organized by San Francisco's "April Coalition," an offshoot of California's Space Now Society. Bringing space enthusiasts together with antinuclear and environmental advocates, the event rejected the strong corporate presence that marked the first Space Day and instead reflected ideas from the Bay Area counterculture. "There is no time to waste uniting the progressive political groups of the United States to wrest control of the national destiny from the militarists," organizers claimed, "Space is the place for the NEW human race!"[123]

By the late 1970s, the prevailing mood of ambivalence, pessimism, and angst over technology writ large that marked the *Limits*-dominated start of the decade had begun to dissipate. Many factors contributed to this including the end of the Vietnam War, the increasing ubiquity of personal computers, and the growing economic importance of high-tech industries in the U.S. economy. In some degree, Gerard O'Neill's visioneering must be counted among these factors. More than anything, O'Neill and the groups he inspired suggested that space settlements and manufacturing could be a new technological frontier where ordinary citizens could play a leading role. The new grassroots pro-technology communities he catalyzed took positions outside the mainstream of traditional technology-oriented organizations, such as professional societies and industrial associations, but their ideas and activism intrigued a growing segment of the American public. In time, enthusiasm for citizen-oriented space exploration helped foment fresh technological visions. These new plans for the future would be based not on launching rockets into space but on manipulating data bits, genes, and molecules.

CHAPTER 4

Omnificent

ıllıllıllıllıllıllıllıllıllıllıllıllıllıllıllıllıllıllıllıllıllıllıllıllıllıllıllıllıllıllıllıllıllıllıllıllıllıllıllıllıllıllıllıllıllıllıllıllıllı

How much would you pay to see the future?
—Advertisement for *Omni* magazine, 1979

Bob Guccione always said he never wanted to be a pornographer. It was just a way to pay for his first love, art. Nonetheless, his X-rated publications made him splendidly rich. Presented as *Playboy*'s raunchier cousin, Guccione's *Penthouse*, with its muckraking journalism and images that left nothing to the imagination, enticed millions of readers each month. The fortune Guccione amassed allowed him to buy a giant Romanesque townhouse in Manhattan and fill it with artworks by Degas, Picasso, El Greco, and other masters.

A flamboyant Sicilian-American, Robert Charles Joseph Edward Sabatini Guccione was born in Brooklyn in 1930 and raised in New Jersey. He considered the priesthood before pursuing an artist's career and traveling around Europe and North Africa.[1] In 1965, he was struggling financially when he met Kathy Keeton, a South Africa–born ballerina turned exotic dancer. "One of the highest-paid strippers in Europe," Keeton caught his eye when he saw her reading the *Financial Times* at the club where she performed. They shared impoverished backgrounds and a dream of "being powerful and living forever."[2] With a small bank loan, Guccione and Keeton—longtime partners but unmarried until 1988—set out to challenge *Playboy*. Launched at the height of the sexual revolution, *Penthouse* sold out its first run in a few days. Four years later, they brought their profitable magazine to the U.S. market.

Buoyed by success and sensing a resurgent interest in science and technology, Guccione and Keeton embarked for new publishing territory. "Bob and I have always been interested in science and science fiction," Keeton explained in 1979.[3] As the first senior editor of their new publication saw it, their goal was to publish "a magazine they wanted to read. For them, it meant no boundaries, any kind of technology that was appealing to them, and it had to be about the future."[4] The lush vehicle they built to explore the future was *Omni* magazine.

Omni's arrival coincided with Americans' "acute need to know more about science and technology," a trend that was becoming apparent in the late 1970s.[5] A magazine profile of astronomer Carl Sagan—his television series *Cosmos* entered production in 1978 and aired in 1980—noticed that years of "ennui" about technology and science had "turned into enthusiasm."[6] The same article also identified Gerard O'Neill as someone who matched "talent for experimentation with a surprising gift for exposition." Dozens of new magazines, television shows, and newspaper sections arrived to meet consumers' demand for information about science and technology. This proliferation represented the largest jump in science-related journalism since the start of the space race.[7]

The emergence of new, articulate spokespeople, such as Sagan and O'Neill, helped stimulate renewed public interest in science and technology. Current events also were a catalyst. "It goes beyond the obvious concerns caused by Three Mile Island, the DC-10, and Skylab," one public affairs expert said. "People realize how much science now influences our daily lives, and how big a role it plays in policy decisions."[8] Shifting demographics in the United States was a third factor. "This generation," said one science magazine editor, "having grown up with Sputnik, the environmental movement, the War on Cancer, the Space Program, and the energy crisis, has come upon an information vacuum."[9] Guccione and Keeton stepped in to help fill this void.

When *Omni* first appeared on newsstands, Bob Guccione promised to show its readers "a future of growing intellectual vitality, expanding dreams and infinite hope."[10] Such a pledge, far removed from previous predictions of societal, environmental, and economic collapse, illustrates perfectly the newly emerging reorientation of public attitudes toward technology. While books like Ehrlich's *Population Bomb*, Mumford's *Pentagon of Power*, and *Limits to Growth* called for reexamining the relations between society, economics, and technology, magazines like *Omni* heralded a "techno-counter-reformation" that presented a more positive view of the technological future.

Part of this newfound optimism stemmed from new sectors of the U.S. economy. By 1980, "biotechnology" and "information technology" had entered the lexicon of investors, journalists, and policy makers. The surge of interest in topics such as genetic engineering and personal computing happened just as American politicians were addressing the fact that many older wells of innovation, especially those associated with traditional heavy manufacturing of products such as cars, appliances, and steel, were drying up. Moreover, new competitors in the global marketplace were supplanting American companies and workers. As a result, industrial and political leaders hoped that biotechnology, microelectronics, and computing technologies would be new sources for America's economic prosperity in the coming decades.

This chapter begins by exploring two strands of technological advances, one in microelectronics and another in molecular biology, that formed what some experts saw as the new double helix of innovation. In the 1970s, basic discoveries in the life sciences led to the emergence of a biotechnology industry while the introduction of the personal computer helped launch a cohort of computer geeks turned business heroes. These new technologies became grand arenas where technically literate entrepreneurs could both play and profit. The tremendous media attention that personal computing and genetic engineering received helped prime the public, business leaders, and politicians for radical claims made later about nanotechnology's potential. Moreover, ideas and research directly drawn from microelectronics and biotechnology proved absolutely essential for early conceptualizations of nanotechnology.

To be sure, O'Neill and his space-oriented visioneering did not suddenly fade away. O'Neill-inspired groups like the L5 Society had started by diligently advocating space as a place that adventurous citizens might one day inhabit. But, by the early 1980s, this goal began to seem more unattainable than ever, and many initial enthusiasts abandoned their utopian aspirations. For example, Stewart Brand, who had helped publicize O'Neill's visioneering, shifted his attention to the idea of virtual communities in cyberspace rather than real ones in outer space.[11] Even O'Neill himself began to devote his energies to more down-to-earth endeavors such as launching high-tech business start-ups. At the same time, the surge of "popular science" publications brought attention to impressive developments happening, not in outer space, but at the levels of the gene, the molecule, and the microcircuit. Magazines like *Omni* helped shift public interest and attention to these marvelous new frontiers. This eventually helped foster ambitions for a new technological future to be realized not in outer space but by controlling and manipulating matter at the level of billionths of a meter—the nanoscale.[12]

Silicon Sorcery

Omni's August 1979 issue included a colorfully illustrated piece called "Wizards of Silicon Valley." It described how, in just a few decades, a strip of land at the end of San Francisco Bay that had once been home to broad expanses of orange groves had blossomed into a potent empire of technological innovation.[13] When the article appeared, the phrase "Silicon Valley," coined by a journalist just eight years earlier, was already becoming synonymous with high-tech vitality.[14] Compared with older and more traditional technological regions such as those around Boston and Detroit, coastal California's aerospace, biotech, and electronics industries seemed to represent the future. Its success-breeds-success culture of entrepreneurialism and risk tolerance appealed to many young people with big dreams. As *Omni* said, Silicon Valley, like medieval Spain had once been, was now the new "launchpad for expeditions into new worlds."[15]

All this—the riches, the inventions, the jobs—owed its existence to the smallest of tools. The transistor, made from a semiconducting element like silicon or germanium, is essentially a tiny device that can switch and amplify electrical signals. Transistors, toggling back and forth, represent the zeros and ones in a modern digital computer and provide the basic logic elements for our modern world. In the years after three scientists at Bell Laboratories in New Jersey invented the first transistor in 1947, a technological mélange of new inventions based on it emerged. As engineers and scientists learned the chemical craft of making transistors and the physics of how they worked, they built ever more complex circuits and devices even as the products designed around them became smaller. Together, the integrated circuit, the microprocessor, and the personal computer came to define the material world of the "Information Age."[16]

Meanwhile, a new way of thinking about technological change emerged. The success of innovative new products helped foster a prevailing belief that future improvements in microelectronics and computing devices were not just desirable but inevitable. This "ideology of the small" coincided with a flurry of popular interest as to what the ultimate limits of miniaturization might be, an enthusiasm that later spilled over to nanotechnology.

In November 1960, for example, *Popular Science* published an article titled "How to Build an Automobile Smaller than This Dot" (an arrow on the magazine page pointed to a tiny black smudge). The author was Richard P. Feynman, a scientist at the California Institute of Technology whose iconoclastic personality, teaching skills, and research acumen made him a legend in the physics community.[17] Feynman based his article on an after-dinner talk he had given a year earlier to the American Physical Society, physicists' professional organization. Feynman's colleagues initially thought that his lecture, puckishly titled "There's Plenty of Room at the Bottom," would poke fun at the growing job market for physicists. To their surprise, the Caltech physicist instead described a world of futuristic possibilities that awaited researchers once they started "manipulating and controlling things on a small scale."[18] To Feynman, "small" transcended the impressive accomplishments already made with transistors and microcircuits, referring instead to the near-atomic realm of molecules, what today we would call the "nanoscale."

In "Plenty of Room," Feynman blended speculative extrapolations of existing technological trends with ideas freely borrowed from biology and science fiction. In the future, Feynman predicted, physicists would be able to synthesize "any chemical substance that the chemist writes down," an ability that might have enormous economic value or might be done "just for fun." However, Feynman himself did no research that built on his over-the-horizon speculations. The 1965 Nobel Prize he shared was for theoretical physics research on how light and matter interact at the most fundamental levels. Although popular magazines like *Time* and the *Saturday Review* reported on the ideas Feynman outlined, his talk quickly sank out of sight. This remained the case for more than three decades until scientists and policy makers rediscovered "Plenty of Room" and interpreted it as visionary portent for what had become known as nanotechnology.[19]

Nonetheless, Feynman's ideas for making things smaller reflected an abiding interest in miniaturization that flourished among Cold War scientists and engineers. As bombers gave way to intercontinental ballistic missiles and the space race accelerated, the ability to build lighter payloads was essential for national security, and ever-smaller electronic and computing devices were key to achieving this. Representatives from high-tech companies, government labs, and the military even created an annual award for small-scale devices that would "extend the frontiers of miniaturization."[20] The increasing sophistication of these gadgets prompted one proponent to predict that the "field of miniaturization knows no bounds. . . . Is it not conceivable that man might someday create a computer as complex yet as small as the human brain?"[21] This was far from the last time futuristic speculations like these would be put forth.

Engineers still had a long way to go, however. When Feynman wrote "Plenty of Room," computers were giant mainframe machines made by large companies like IBM and General Electric. As electrical engineers combined a few, then dozens, and then hundreds of transistors into more compact and complex devices, a "tyranny of numbers" presented them with a considerable obstacle. Adding just a few more components to a circuit meant that several new interconnections, laboriously fashioned via tiny wires by technicians using tweezers, were needed.[22] Adding more transistors, in other words, made a device's complexity rise exponentially. Meanwhile, if a single connection failed from harsh weather or a bumpy ride into space, it could render an entire system useless.

Engineers began to imagine the possibilities if they fashioned the entire electronic circuit, including the wires and transistors, onto a single semiconducting wafer. Although not fundamentally different in principle from hand-wired microcircuits, these "integrated circuits" promised lower costs and simplified designs. As they had done for transistors, deep-pocketed military and aerospace contractors helped drive integrated circuit technologies forward.[23] Innovative companies like Fairchild Semiconductor, founded in 1957 near San Jose, made a fortune mass-producing microcircuits. Between 1961 and 1965, sales of integrated circuits increased from $500,000 to more than $50 million.[24] Leading figures in the electronics industry, looking at their balance sheets and factory floors, began to see a pattern. "The technology improves," remarked one Westinghouse executive, and yet "the cost drops."[25]

In 1965, the trade journal *Electronics* asked Gordon E. Moore to speculate about the "future of microelectronics" for its April issue. Moore was well placed to offer such a perspective. "One of the new breed of electronic engineers," Moore possessed a technical mastery of semiconductor chemistry and solid-state physics.[26] With a doctorate from Caltech, he was one of the original engineers who started Fairchild Semiconductor, a seminal Bay Area firm, and he directed the company's research and development efforts.

Moore candidly titled his essay "Cramming More Components onto Integrated Circuits." But it was as much about economics as it was about electronics. Using just five data points, Moore graphed the number of components on a typical integrated circuit versus time. His plot, when put on a logarithmic scale, fell on a nearly straight line that, when extrapolated into the 1970s, shot upward. In other words, Moore predicted a doubling of computer performance, as measured by the number of transistors on a single chip, taking place almost every twelve months.[27] Maintaining this technological trajectory required the fabrication of ever-smaller components, a feat engineers at electronics companies readily and reliably achieved.

What became known as "Moore's Law" initially found slight recognition outside the confines of the electronics industry. But Moore's carefully qualified observations gradually morphed into a general statement of exponentially increasing technological power. Industry executives, economists, and engineers invested great faith in its metronomic pace and used Moore's Law to make long-range industrial plans. The steady improvement in computer chip performance, achieved even as costs plummeted, became the heartbeat of global industries.[28] Applied originally to the relatively esoteric subject of circuit complexity, Moore's Law itself seemed to grow exponentially in popularity among engineers, economists, and journalists, who began citing it as a descriptor for technological progress that could be applied to other areas such as telecommunications and genetics.

The success of companies like Fairchild Semiconductor helped the region around Stanford University and San Jose become a technological powerhouse. More than a hundred businesses spun off from Fairchild in the 1960s as an entire cohort of entrepreneurial engineers started their own companies. Of these, the most important venture was one Moore and his partner Robert Noyce launched themselves in the summer of 1968. Intel Corporation (short for "integrated electronics") initially focused on making circuits for computer memory, a market the company soon dominated. Within a few years, Intel's engineers branched out. Using instructions stored in their electronic memories, the new "microprocessors" Intel introduced could perform several functions. Their success depended on engineers' abilities to take common materials like silicon and transform them into complex chips, circuits, and, eventually, profitable consumer goods. Engineers' fabrication techniques soon became "so precise that people in Silicon Valley" spoke of "counting atoms."[29]

Intel, of course, was just one of many companies thriving in the fecund Cold War technological ecosystems found in coastal California, around Dallas and Atlanta, and along Massachusetts's Route 128. New high-tech companies, especially those in Silicon

Valley, began to receive a growing amount of attention. This was especially the case after a cohort of computer hobbyists turned business leaders launched successful new computer and software companies. Meanwhile, established companies watched profits soar. "The basic thing that drives the technology is the desire to make money," one executive said. In 1969, Intel sold only $600,000 worth of products. Ten years later, it earned a phenomenal $78 million from revenues worth over $650 million.[30] Numbers like these made investors and politicians pay attention. Journalists certainly did. Between 1977 and 1980, for example, the *New York Times* and the *Los Angeles Times* ran more than 120 articles on Silicon Valley's entrepreneurial activities, often highlighting the profits that future-looking venture capitalists were collecting on their investments.[31]

Moore's Law was fundamentally about what computers could do. But it said very little about what people would do with them and how they would do it. One sense of the possible came from Douglas C. Engelbart, an electrical engineer working at the Stanford Research Institute, just a short drive from Moore's corporate laboratory. Around 1962, Engelbart had several profound insights about how people would interact with digital computers in the future. He anticipated that engineers would give these new machines an interactive capacity for "augmenting human intellect."[32] In other words, instead of just processing data, computers could help people think, reason, analyze, and communicate.

Over the next five years, Engelbart designed and built a host of novel devices that changed the interaction between people and computing machines. Engelbart linked his engineering activities to a wish to help people, as he later explained, "cope better with complexity and urgency and the problems of the world." For more than two decades, he worked assiduously to achieve his goal of interactive computing.[33] By combining expertise in computer engineering with a vision for how the technologies he advocated would shape the future and then actively promoting it, Engelbart worked as a visioneer.

In December 1968, as the performance of computer chips maintained the steady rate of improvement Moore envisioned, Engel-

bart debuted his research at a computer conference in San Francisco. In front of an auditorium packed with people eager to see
the "augmented knowledge workshop," Engelbart introduced nov-
elties such as opening files with a "mouse," electronic mail, and
working in real time with multiple "windows" on a screen.[34] Today,
we take these tools for granted. But, in 1968, they represented radical, even heretical, ways for people to interact with computers in
real time instead of the traditional mode of batch processing via
stacks of punched cards.

Although the commercial computer world paid little notice,
many witnesses to Engelbart's "mother of all demos" were blown
away. They, in turn, went off to develop their own ideas and tools
for improving computer-human interaction. Engelbart's visioneering instantiated a particular view of a technological future. It suggested what was possible and also helped motivate other computer
experts to explore further and see if such possibilities could be realized. As Alan Kay, a computer scientist sympathetic to Engelbart's ideas, said in 1971, "The best way to predict the future is to
invent it."[35]

Engelbart's ideas, even though they were slow to find commercial acceptance, helped catalyze a shift in people's perception of
computer interaction. In a widely read *Rolling Stone* article, Stewart Brand, who helped organize Engelbart's visionary computer
demonstration, said, "Ready or not, computers are coming to the
people. That's good news, maybe the best since psychedelics."[36]
What emerged was a grassroots "computers for the people" movement that coincided with the efflorescence of citizen interest in
O'Neill's humanization of space. Membership in one group sometimes spilled over to the other, and these enthusiasts were some of
the first people interested in nascent ideas for nanotechnology.

An essential spark for this was the appearance of the first personal computers. In 1974, for example, former amateur rocket enthusiasts operating a small company in New Mexico began selling
the Altair 8800. What *Popular Electronics* trumpeted as "the
world's first minicomputer kit" was sold via mail order for $379.[37]
It hardly appeared revolutionary. Only slightly larger than a
toaster, the Altair had neither screen nor keyboard. Instead, its

owner flicked switches on the Altair's front panel to program the machine. A user doing this correctly was rewarded by a sequence of flashing lights. Despite its modest capability, thousands of hobbyists bought and built Altair kits. In places like Palo Alto, where now-legendary groups like the Homebrew Computer Club met, computer aficionados, hackers, and future computer executives gathered to swap gear and share ideas. "Virtually every city in the country now has its own band of computer hobbyists," *Omni* reported in 1978.[38] Some of these hobbyists changed the face of American industry. Bill Gates and Paul Allen, for instance, got their start writing software for the Altair before founding what was originally called Micro-Soft.

In addition to hardware and algorithms, a set of social principles helped foster the first wave of the personal computing movement. One of its early leading figures was Theodor Holm (Ted) Nelson, the financially independent son of Oscar-winning actress Celeste Holm. Born in 1937, Nelson grew up with his grandparents in Greenwich Village and later earned a master's degree in sociology from Harvard. In 1974, Nelson, an autodidact when it came to computer science, self-published a book called *Computer Lib/Dream Machines* (it was actually two books bound together, Janus-like, into a single volume). Inspired by the *Whole Earth Catalog*'s cut-and-paste style, Nelson combined technical descriptions with a vision for the future of computing interspersed with his own irreverent opinions ("Down with Cybercrud!"). Nelson believed that simple and affordable computers could foster personal freedom for ordinary people instead of being oppressive tools of a technological elite.[39] *Computer Lib*'s cover made Nelson's vision unambiguous—a fist raised, Black Panther style, in front of the hated IBM punch card, with the decree "You can and must understand computers NOW."[40]

Throughout the 1970s, computer aficionados shared, borrowed, and stole copies of *Computer Lib/Dream Machines*. Not long after Nelson's opus appeared, two long-haired hobbyists working in a Los Altos garage began to hand-assemble personal computers. The name that Steve Wozniak and Steve Jobs gave their machine— "Apple"—smacked of the forbidden fruit Nelson had envisaged.

Mark Miller, Eric Drexler's friend and fellow space settlement advocate, was so taken with the vision in Nelson's book that he sought out its author in hopes of helping bring it to fruition.

Ted Nelson revisited his hopes for "computer lib" four years later with an essay, lavishly illustrated with artful images of circuits and microprocessors, in *Omni*. Personal computers were already becoming ubiquitous in offices and living rooms. "The revolution has begun," Nelson wrote. "Computer Lib has become a fact."[41] By the time this issue of *Omni* hit the newsstands, more than two decades of technical developments had matured into a potent and pervasive ideology. It intimated that smaller didn't translate only to better and cheaper consumer electronics. Smaller was also an expected, perhaps even inevitable, attribute that implied incessant technological change for the better. Over time, this faith migrated to other areas. Advocates of nanotechnology found a template for their predictions in the history of the microelectronics industry as well as their own professional experiences.[42] Drawing from the message of Moore's Law, they forecast a technological future based on mastery of matter at the smallest scales, a future that would be profitable, predictable, and positive.[43]

Splice of Life

His hands holding a chilled beaker, a "scientist clad in white gently spools spaghetti-like threads of DNA onto a glass rod." He then treats it with "enzymes that will clip away all but a chosen gene, then inserts it into the genetic material of a bacterium." When the operation is over, the microbe will possess "powers that nature never gave it."[44] This was how *Omni* started a feature article on biotechnology in March 1980. The infectious excitement it conveyed encompassed both the field's new technological opportunities as well as fortunes awaiting the "fledgling firms willing to risk their money."

In scores of articles, *Omni* and other popular magazines described how advances in areas such as neuroscience, cloning, and genetic engineering were transforming the life sciences. Researchers' abilities to study the structure and function of proteins and DNA increased remarkably during the 1960s.[45] A decade later, molecular biologists could directly manipulate and modify these biomolecules. What began as attempts to answer basic biological questions transformed into an ensemble of technologies with an array of practical applications.

These headlines came with the scent of profits and not a little peril that helped amplify interest among curious citizens, investors, and politicians. One common trope was the idea that microbes like *E. coli* bacteria could be genetically transformed into "DNA factories" that would manufacture medicine and other useful chemicals.[46] At the same time, scientists and activists expressed concerns that these genetically modified organisms could escape from labs and proliferate unchecked, wreaking environmental havoc.[47] As was the case with microelectronics, advances in molecular biology depended on researchers' ability to understand and precisely control matter on a very small scale.

Researchers in one field occasionally looked to the other for insights and useful analogies. For example, the discovery in the early 1960s that proteins can switch between different states depending on the presence of a specific chemical suggested a comparison to electrical switches such as transistors.[48] In fact, in 1965, when the Royal Swedish Academy of Sciences presented Nobel Prizes to Jacques Monod and François Jacob for their research on the genetic control of enzymes, it analogized their work to recent achievements in electronics. The Swedish Academy noted that each living cell "contains hundreds of thousands of chemical control circuits, exactly harmonized and functioning infallibly. It is hardly possible to improve on miniaturization further."[49] This would by no means be the last time that researchers or journalists likened the behavior of biomolecules to electronic components like transistors and imagined constructing computers and other devices from them.

Proteins were not the only macromolecules that biologists were exploring. As James Watson and Francis Crick famously showed in 1953, a molecule of deoxyribonucleic acid forms a double helix. The portion of DNA's elegant structure that has the "information" for making a specific protein is called a gene. It became common-place for scientists to imagine DNA as a molecule "programmed" like a computer to transmit a "code."[50] Molecular biology became intertwined in its ideas and metaphors with fields like cybernetics and computer science. One of the most pressing research problems for molecular biologists became: how does the information stored in DNA's double helix get translated into proteins?

There were some hints. A number of scientists, for example, studied the possibility that ribonucleic acid (RNA), which is similar to DNA but made from just a single strand of nucleotides, acted as an intermediary between DNA and protein making. In May 1961, using *E. coli* bacteria as a model specimen, researchers at the National Institutes of Health in Bethesda, Maryland, created a synthetic RNA molecule using just uracil, an elementary nucleotide. After inserting this "messenger RNA" into *E. coli*, they saw that it stimulated the production of a specific amino acid called phenylala-nine. The press framed this discovery as a first step toward deciphering the genetic code, and the news rivaled the launch of the first Soviet and American astronauts as the year's top science story.

During the 1960s, scientists increasingly referred to biomole-cules like DNA and RNA as systems for information storage, retrieval, and replication. Popular accountings promulgated the idea that an organism's cells functioned like miniature machines, following instructions their molecular "programs" transmitted. A 1962 article in the *San Francisco Chronicle*, for instance, compared the function of RNA to the computer punch cards that controlled machines on automobile assembly lines. Marshall Niren-berg, who shared a Nobel Prize in 1968 for his research on the genetic code and protein synthesis, even likened "messenger RNA" to a robot that will "faithfully carry out the instructions" given to it by DNA. By using "elementary messages" written in the language of biochemistry, people were developing the ability to communicate "directly with the robot."[51]

Analogizing DNA and RNA to computer codes and robots raised an important issue. Computers, of course, could be reprogrammed to do many things. And, if one imagined biological cells as information-processing machines *and* if it was possible to control the information they received, then it was logical to imagine that scientists could eventually engineer new forms of life. But, prior to the late 1960s, much of the discussion was based on what it *might* be possible to do. Within a few short years, however, practicable "genetic engineering" became both an increasingly common phrase in Americans' lexicon and a topic of ethical debate among scientists.[52] What made this possible, and helped create the modern biotechnology industry in the process, was recombinant DNA (rDNA) technology.

The ability to remove a snip of genetic material from one organism and combine it with DNA from another organism was a goal eagerly pursued at molecular biology labs across the United States. California's Bay Area was one major center for this research. In the early 1970s, teams led by Stanley Cohen and Herbert Boyer at Stanford and the San Francisco campus of the University of California began DNA-splicing experiments. In October 1972, Stanford's Paul Berg announced that his lab had successfully combined DNA taken from a bacterial virus with mammalian genetic material (in this case, a monkey tumor virus). For the first time, DNA from two different species had been combined. The next step was to try similar techniques with organisms more complicated than viruses. Within two years, researchers extracted DNA from an African clawed frog and successfully introduced it into an *E. coli* bacterium. Once it was copied, the frog DNA could be transcribed into RNA, a key step toward synthesizing proteins. This led to speculation that scientists might be able to go even further and design entirely new proteins.[53]

News reports about this "genetic alchemy" highlighted the fact that the new technology offered scientific and economic rewards as well as significant risks. Years earlier, one Nobelist had warned that "tampering with life" might create a future more dangerous "than that implied in the instruments of mass destruction."[54] Now that time seemed to have arrived.

In February 1975, an international group of prominent molecular biologists met at the Asilomar Conference Center on California's Monterey Peninsula to formulate a plan for addressing technical issues of safety—moral and ethical issues took a backseat—in labs doing rDNA research. One worry scientists expressed was that modified bacteria might become a pathway for transmitting cancer or other diseases (an appropriate worry, since early DNA-splicing experiments involved a monkey tumor virus and *E. coli*, a bacterium common to the human intestinal tract). Without strict laboratory controls, new microorganisms created via rDNA techniques might escape, replicate unchecked, and become an environmental pollutant or even a biological hazard. Between 1975 and 1979, in fact, more than half a dozen communities passed resolutions to govern rDNA research.[55]

Capitalism trumped caution, however. Over the next decade, scores of new biotechnology companies joined established electronics firms to become a vital economic force in tech-heavy regions like the Bay Area and around Boston. One article, showing the length that the *Limits to Growth* shadow still stretched, suggested that this "genetic gold" might "relieve Malthusian anxieties about a future without sufficient raw materials."[56] Decades of government funding stood poised to combine with venture capital to create expanded investment and manufacturing opportunities. Bacteria, journalists said, might become the future's new "factories," churning out insulin or other valuable biomolecules.[57]

Factories suggested profits. J. D. Bernal and Freeman Dyson had imagined altering the Flesh. Now entrepreneurial molecular biologists and their corporate backers sought to get rich doing it. In 1972, in fact, a biochemist employed by General Electric filed for patent protection on a bacterium he had engineered through selective breeding to consume oil spills. New rDNA techniques appeared to offer an even easier way to create useful new organisms. In November 1974, Stanford and the University of California filed a patent application for Cohen and Boyer's DNA manipulation process.[58] Two years later, Boyer helped found one of the first biotech start-ups. Genentech, short for "genetic engineering technology," soon announced that it had genetically engineered bacteria that could produce human insulin. The potential commercial applications overrode any inclination of Congress to regulate the new technology and, in 1980, the U.S. Supreme Court ruled that new forms of engineered life engineered could be patented.[59]

By 1980, investors and stockbrokers were already bullish on what financial analysts had branded "biotechnology" and the *Economist* called "frenetic engineering."[60] When a major brokerage firm sponsored a workshop it called "The Genetic Revolution," its organizer imagined that only a few dozen people would attend. Instead, hundreds of eager investors showed up. Interferons, a class of proteins that interfere with certain viral infections and might help fight cancer were of particular interest. Investors' excitement about the biomolecule—one researcher described interferon as "a substance you rub on stockholders"—amplified scientists' interest. As one journalist later noted, the overselling of interferon was a "dress rehearsal for bio-technology" of all types, an experience that left the impression that "image meant more than substance."[61] As we'll see, something quite similar happened two decades later with nanotechnology, both in terms of sometimes overblown anticipation as well as the fact that commercial interests trumped concerns about potential hazards the technology might pose.

Brokerage firms such as E. F. Hutton estimated the annual global market for chemical "magic bullets" like interferon could be $100 million or more. Following its public stock offering in October 1980, Genentech's share prices doubled in one day. Five months later, Cetus, another Bay Area biotech firm, raised $120 million in one day with the largest initial public stock offering to date in U.S. corporate history.[62] A resurgent pro-business environment—relaxed federal laws that encouraged high-tech investing, lower taxes on capital gains, and legislation designed to foster the commercialization of academic research—helped drive this "biomania."

Investors and venture capitalists also eagerly placed bets on biotechnology because of their past experiences with microelectronics and other information technologies. Ronald Cape, who cofounded Cetus in 1971, told *Omni* that the "large companies have been caught flat-footed by these rapid developments in biology. The same was true for the semiconductor field." As Cape saw it, risk-taking biotech ventures were the "upstarts that came in to fill the vacuum."[63] Similar patterns of expectation and anticipation would reappear in the dot-com boom of the 1990s and then again with nanotechnology. Experts predicted these new technologies would

generate a new manufacturing base for "postindustrial America."
It was not unusual to hear scientists and journalists speak of the
potential social changes the genetics or personal computer revolutions might bring in the same terms that their counterparts had
used with regard to nuclear power or space satellites a few decades
earlier. In the midst of this new burst of enthusiasm for the technological future, new magazines like *Omni* appeared ready to both
explain and entertain.

"Technologists of the World, Unite!"

If *Omni* showed the future, it was going to look stunning. Guccione spared little expense, and *Omni* looked like no other magazine in print. Full-color illustrations filled entire pages, and each
issue featured gorgeous photographic essays. Guccione, ever the
art lover, made sure that *Omni* presented evocative and enigmatic
artwork by new and established artists. Guccione and Keeton even
developed a special font for the titles of articles and columns, which
added to *Omni*'s sleek, futuristic look. Their attention to appearance paid off. Philip Handler, president of the National Academy
of Sciences, praised *Omni* as "handsome and attractive," both "es-
thetically and intellectually."[64]

Omni's work environment reflected Guccione's experience with
other publishing ventures. Its first senior editor, Trudy E. Bell,
came to *Omni* after working at the staid and lofty *Scientific American*. "It was this absolutely frenetic environment," she recalled.
"Guccione and Keeton were very inspirational people who would
catch you up in their imaginative vision.... They could almost
warp reality."[65] A visitor to *Omni*'s first office on Third Avenue
was likely to exit the elevator and find a photo of a naked young
woman sewing stars on the American flag à la Betsy Ross.[66] Despite feminists' loathing of *Penthouse*, Guccione staffed the top
tiers of his magazines with women. Keeton played a large role in
helping run *Omni*'s daily activities. Guccione, meanwhile, preferred to host hours-long editorial meetings at his opulent mansion on the Upper East Side.

"How much would you pay to see the future?" *Omni* asked its readers in an early campaign to increase circulation. Quite a lot, it turned out. The magazine's cover price, $2, was one of the highest for the time, yet *Omni* soon could boast that it sold close to a million copies every month. By 1980, Keeton claimed some four million people read it monthly. *Omni* had especially strong appeal for the classic baby boomer demographic: eighteen- to thirty-four-year olds, mostly but not entirely male, and single. According to one executive editor, "We liked to think of our typical reader as someone who read the Sunday *New York Times*. Up-scale, financially and educationally."[67] A large fraction were recent college graduates, some perhaps starting new jobs in high-tech industries, with considerable disposable income.

Where earlier publications like *CoEvolution Quarterly* and the *Whole Earth Catalog* had stressed frugality and pragmatism in their technology-oriented consumerism, *Omni* presented a hedonistic view of a future made shiny and sexy by sophisticated technology. Instead of pitches for self-composting toilets and wood stoves, one found advertisements for home electronics, high-fidelity audio gear, and European sports cars. These were sprinkled amidst stories about personal ultralight aircraft, space tourism, and life-extension research. While certainly not the sort of consumerism that *Whole Earth* readers would have necessarily approved of, both publications offered images of a desirable lifestyle fashioned by technology with a bit of escapism thrown in.[68] But where *Whole Earth* reflected readers' aspirations for individual self-sufficiency, *Omni* presented the views of technology enthusiasts who saw corporate-based innovation as a component of their own personal fulfillment and financial success.

Omni's overall advertising strategy reflected its target demographic and its aspirations. Any concerns as to whether *Omni*'s link to the soft-core porn industry might unnerve main street advertisers proved unfounded. Supermarkets carried *Omni*, airlines offered it for in-flight reading, and the magazine broke records for advertising revenues. Advertising fees—up to $9,000 for a full-page color spread—ensured that *Omni* could afford to hire brand-name writers and present lavish pictorial layouts.

Many of the companies who bought space in *Omni* produced advertisements designed to appeal to high-tech enthusiasts and future-oriented readers. For example, Champion International, a traditional wood and paper company, ran a series of high-profile advertisements around the theme of "The Future is Coming." One of the company's ads featured a pair of "healthy, hearty, 'Methuselahs' " and discussed a variety of life-extension options ranging from organ transplants to cryonic preservation (perhaps counterbalancing *Omni*'s pervasive pitches for cigarettes and top-shelf liquors).[69] Less expensive ads in the magazine's back pages ran a wide gamut: pitches for Mensa, Dianetics, and the Rosicrucians shared space with ones for solar-powered calculators and home telescopes. *Omni*'s ads also suggested its readers possessed a modicum of technical skills. The Heath Company, whose popular kits electronic hobbyists could build at home, advertised in *Omni*, for example. Out of this mélange, we can construct an image of a typical *Omni* reader: a sophisticated, tech-savvy, and open-minded person, probably male; comfortable with computers, intrigued by space exploration and biotechnology; interested in both the "good life" of high-end toys and expensive liquor as well as the latest news from the borderlands of "science-faction."[70]

Omni was by no means the only magazine of its time that combined nonfiction essays with a healthy dose of fiction and fantasy. For example, *Future Life* debuted its premier issue in April 1978. A sister publication of *Starlog*, a science fiction magazine, *Future Life* covered topics similar to those in *Omni*. Carolyn Henson wrote a regular column for *Future Life* about the O'Neill-inspired space movement, for instance. However, when it came to style, circulation, and top-tier advertisers, *Omni* was light-years ahead.

Omni promised readers a "ticket to infinity" with its diverse nonfiction content.[71] The topics Guccione and Keeton chose reflected their interest in all kinds of futuristic technologies including research that straddled the border of fiction and fact. Deeply curious about the paranormal, Guccione and Keeton made sure that articles about psychics, the Loch Ness Monster, and UFOs appeared monthly in a section, printed on bright red pages, they called Antimatter. Scientific research on life extension especially piqued their interest. *Omni*'s first issue featured an article called "Some of Us May Never Die," and the magazine often reported on cryonics, in which people are frozen at their time of death in the hope that future medical technologies might revive them. Guccione and Keeton's interest in living longer wasn't a mere curiosity. In the mid-1980s, they launched a new magazine called *Longevity* that promoted both healthy living and life extension.[72]

Omni's attention to science and technology situated outside the mainstream was understandable. In the late 1970s, Americans, including many reputable scientists, expressed tremendous interest in "New Age science," Eastern mysticism, and parapsychology.[73] Pop culture reflected this curiosity. Starting in 1976, Leonard Nimoy, famed for portraying Mr. Spock in the *Star Trek* series, hosted *In Search Of. . . .* Over a typical season, the popular television show mixed stories about witch doctors and Bigfoot with ones about cryonics and cloning. *Omni*'s paranormal coverage, like that of *In Search Of . . .* , reflected lingering public skepticism about scientific "truths" produced by the mainstream research Establishment. It also suggested readers' desire for alternative ways of thinking about nature, especially when there existed any hint of conspiracy theory or government cover-ups.

Guccione and Keeton's curiosity about "fringe" topics also reflected their own willingness to flout mainstream convention. Pornography itself hovered at the fringe of the publishing and movie business in the late 1970s (even as he was launching *Omni*, Guccione was producing the film *Caligula*, a multimillion-dollar epic that *Newsweek* called a "cavalcade of depravity"). *Omni*, to be sure, never fully shed its pedigree, and its cartoons and artwork maintained a slightly titillating sensibility. Nevertheless, *Omni*'s ability to pull in mainstream corporate advertisers proved it had overcome skeptics' initial judgment that it was going to be "*Penthouse* in space . . . a version of *Nature* featuring cartoons of three-breasted women."[74]

Guccione and Keeton, of course, also solicited plenty of articles for *Omni* about cutting-edge technologies and the latest scientific research that had no links to the paranormal. Each issue's silver-paged section called Continuum summarized noteworthy research and intriguing technological developments. Biotechnology, especially as it related to genetic engineering and cloning, appealed to readers as did articles about the science of sex. Stories about computers and high-tech electronic gadgets—their performance and affordability enhanced by the steady beat of Moore's Law—paralleled *Omni*'s growing advertising revenue from computer companies such as IBM and Apple. Interviews with unconventional or iconoclastic researchers were a regular feature as well, and notables such as Freeman Dyson, John Lilly (known for his studies of dolphin-human communication), futurist Alvin Toffler, and physicist Richard Feynman all appeared in *Omni*'s first year.

Omni's fiction proved another big attraction for readers. Guccione recruited veteran science fiction author and editor Ben Bova to *Omni* in late 1978. Bova, who had edited *Analog* (America's preeminent science fiction magazine) for several years, saw *Omni* as "the magazine I had dreamed of since childhood . . . a slick, big, gorgeous magazine about the future." He took the view that *Omni* wasn't about science per se, which people "thought of as spinach. It doesn't taste good but it's good for you." Instead, "*Omni* was about the future. It's lemon meringue pie."[75]

By 1980, Guccione had promoted Bova to the position of *Omni*'s editor in chief. From this vantage he attracted a cohort of writers, some young and others well established, who wrote compelling fiction without resorting to worn sci-fi tropes. Several of them became award-winning authors after appearing in *Omni*. Bova wanted to bring sci-fi writing to a broader audience, and many of *Omni*'s fictional pieces addressed the same technological topics—space, bioenhancements, computers, robotics, and so forth—that appeared in its nonfiction essays.

For a magazine based on the premise of technological optimism, *Omni*'s fiction could be remarkably dystopian. Before William Gibson published his 1984 tech-noir classic *Neuromancer*, he introduced *Omni*'s readers to then-exotic concepts such as cyberspace and virtual reality. In his story "Johnny Mnemonic," data traffickers with computerized brain implants evade biologically modified assassins moving through an urban underworld controlled by corporate Japanese yakuza syndicates. Orson Scott

Card, later renowned for his Ender's Game series, contributed an even more unsettling tale. "A Thousand Deaths" takes place after a complacent and decadent United States has been taken over by Russia. Convicted of subversion, playwright Jerry Crove is condemned to die, over and over. Each time, interrogators upload his memories to a new cloned body and execute him again. Failing to rehabilitate Crove, the Russians send him and hundreds of other dissidents, their biological functions suspended, "upward endlessly into the stars" to colonize a new and presumably freer world.[76]

Omni consistently exposed its readers to new ideas about what the technological future might be like. But whether it was fiction or nonfiction, the subject that *Omni* covered most often in its first few years was space exploration. Guccione, Keeton (who agreed to join L5's board of governors), and Bova were all staunch supporters of an expanded human presence in space. The pages of *Omni* reflected their pro-space agenda with stories about NASA missions and the successes of the Soviet space program. Meanwhile, full-page advertisements from industry giants like Rockwell International, which built the space shuttle, helped pay *Omni*'s bills.

Gerard O'Neill's visioneering got especial attention in *Omni*. Almost every issue between 1978 and 1982 mentioned space colonies, satellite solar power stations, space-based manufacturing, or the growing pro-space movement. *Omni*'s monthly editorials routinely made the case for a more robust American space program executed in the O'Neill style. Besides offering free advertisements for the L5 Society, *Omni* helped promote O'Neill's *High Frontier* with laudatory reviews and frequent mentions. *Omni* also shared its mailing list of subscribers with O'Neill's Space Studies Institute, while Bova himself wrote both a novel (*Colony*, published in 1978) and a nonfiction book (1981's *The High Road*) that riffed on O'Neill's concepts.

Omni's depictions of O'Neill would have resonated with its readers. In stories and interviews, O'Neill was described as a handsome Ivy League physicist-entrepreneur with humanistic leanings and a beautiful wife. When not working on large-scale research projects, O'Neill piloted his own plane to attend NASA meetings or testify before Congress. He also expressed a passion for using technology to make both societal improvements and some money.[77] In *Omni*'s presentation, O'Neill was indeed a visioneer, a future-looking scientist who promoted bold new technological ventures that would create worlds of new possibilities.

Reflecting the grassroots pro-space movement itself, *Omni*'s political messages defy easy characterization. Like the eclectic, freewheeling assemblage of Stewart Brand's *CoEvolution Quarterly*, *Omni*'s essays and especially its fiction expressed shades of social liberalism and antitotalitarianism. But *Omni* valorized free markets and conspicuous consumerism while still endorsing megaprojects for energy and space whose success would require government support, a contradiction readers let pass. Well before Albert Gore Jr.'s well-publicized support of the Internet, *Omni* opined that the Democratic senator's pro-innovation perspective would make him a good presidential candidate. A few years later, Gore wrote an opinion piece for *Omni* that blasted Reagan-era "supply-side science," in which short-term economic goals trumped research and long-range investments. However, after Reagan's 1980 election, Bova had cheered that "the sunshine is returning" after years of stagnation and doomsday predictions.[78] Yet dangers still existed: the Soviets threatened domination from space and in military arenas; the Japanese were taking a leading position in the high-tech marketplace. However, sunshine was still returning: *Omni* cheered innovations in robotics, space commercialization, biotech, and electronics that could usher in a "new economic upsurge."

Guccione and Keeton wove environmental concerns into their technological enthusiasm. Kenneth Brower, *The Starship and the Canoe*'s author, contributed columns on ecological issues, while articles on alternative energy sources appeared, as did pieces opposing whaling and promoting "ecoshelters." Guccione himself was a huge booster of nuclear fusion as a source of energy—joining light elements together as opposed to splitting heavy ones like uranium apart and, in theory, safely generating clean nuclear power. *Omni* devoted an entire issue to the topic in 1981, and Guccione invested some $17 million of his own fortune in a private fusion energy venture that, at its peak, employed more than eighty people.[79]

Overall, *Omni* tended toward a viewpoint that was moderately libertarian by early 1980s standards. Its articles and opinion pieces aligned with the "less government–lower taxes" ideology that became increasingly prominent during the Reagan era. This preference for free enterprise extended into outer space. In 1979, *Omni* joined the L5 Society to oppose the United Nations "Moon Treaty." If ratified, the agreement would have turned jurisdiction of celestial bodies, including the moon and asteroids, over to the international community. Many space advocates opposed the plan, thinking it would discourage private enterprise, and L5ers like Carolyn Henson, James Bennett, and Eric Drexler successfully lobbied U.S. politicians. *Omni*, meanwhile, published opinion pieces saying the treaty was "carefully designed by Communist and Third World parties to discourage private industry's role."[80]

Reaganesque pro-market themes appeared elsewhere in *Omni*. One writer profiled "corporate titan, philanthropist and futurist" Malcolm Forbes. Another *Omni* correspondent reported from a national innovation symposium attended by experts from Silicon Valley and soon-to-be Rust Belt cities. She noticed that many participants wore ties decorated with images of Adam Smith. The design, she concluded, "does for economists what the Lacoste alligator does for golfers" and was a personal favorite of Milton Friedman, a Nobel Prize–winning economist who extolled the virtues of free markets.[81]

References to *The Limits to Growth* remained a reliable shorthand for any sort of restrictive set of ideas about the future. "We don't want to ruin the earth. We just want to get off it," one self-identified "progressmaniac" wrote in *Omni*, "go our elitist ways, splicing genes, and meddling with human evolution. . . . And, as the Schumachers [economist E. F. Schumacher] of the world churn out their new improved rice paddies . . . we'll be speeding toward the stars. Technologists of the world, unite! You have nothing to lose but your environmental impact statements."[82]

Guccione and Keeton's technological enthusiasm, combined with the opinions and topics that filled *Omni*'s pages, foreshadowed viewpoints found a decade later in cyber/computer-oriented publications such as *Mondo 2000* and *Wired*. With a wider circulation and more consumerist orientation, *Omni* was never as overt in its politics as *Wired*, which, during the Clinton years, assumed a messianic tone about the technological future. This "California ideology," as critics branded it, "promiscuously combines the free-wheeling spirit of the hippies and the entrepreneurial zeal of the yuppies" and it was seen largely as a 1990s-era phenomenon.[83]

Nonetheless, faint but clear traces of this techno-libertarianism can be found a decade or more earlier in the pages of *Omni*, as can many of the "high tech intellectuals" and New Right politicians who later promoted computers, the Internet, and the dot-com boom.[84] For example, in 1980, Newt Gingrich was a newly elected Republican congressman from Georgia keen to maintain America's technological superiority. In a letter to *Omni*, Gingrich passionately argued that a revitalized space program was key to America's future.[85] He was less clear as to whether this should be achieved via O'Neill-style "space populism" or massive new government programs. Advised by futurists like Alvin Toffler and Isaac Asimov, Gingrich expressed similar ideas in *Window of Opportunity*, his "blueprint for the future" that Tor Books, better known for its science fiction and fantasy titles, published in 1984.[86] A decade later, *Wired* applauded Gingrich as a cyber-revolutionary.

Offering viewpoints from rising political stars helped *Omni* stand out among the overall surge of media interest in science and technology. Between 1978 and 1984, fifteen new science and technology magazines rolled off the presses. In addition, more than thirty newspaper sections and television shows devoted to "popular science" appeared.[87] Bubbles collapse, of course, and by the late 1980s many of these high-profile science magazines had folded, a decade or less after they first came out. Nonetheless, *Omni* survived the bust and continued to attract hundreds of thousands of readers. It maintained a readership because the writing in *Omni* was not just about "science qua science" but touched also on politics, policy, and economics with a large dollop of the paranormal added to the mix.[88]

In fact, *Omni* kept its format largely unchanged until print publication ended in 1996. By the time the magazine expired, its style, with covers still featuring sleek rockets and spacescapes, looked increasingly out of touch with the new cyber-times. Keeton experimented briefly with an online version of *Omni* but, by this point, the *Penthouse* empire was struggling against tax bills and a rising tide of cheap Internet pornography. Keeton's death in 1997 at age fifty-eight not only dealt Guccione a deep personal blow but also ended the *Omni* era.

As a venue for popularizing science ("science for the people") and a vehicle for popular science ("science of the people") *Omni* succeeded.[89] However, it wasn't a tool for science education in the same manner as *Scientific American* or *National Geographic*. Instead, *Omni* helped create a different kind of "scientific American," one who sought to learn about science and technology as much for the sake of entertainment as for edification. With *Omni*, Guccione and Keeton broke new ground in how the media presented technology and the future. Their writers and editorial staff, along with the scientists, engineers, and visionaries they wrote about, served as nodes in a larger network that presented and popularized a more optimistic view of the future and its technologies. In 1986, when the first major article popularizing the revolutionary potential of nanotechnology appeared, it was—where else?—in *Omni*.

The Right-(Wing) Stuff

Starting in the late 1970s, readers of *Omni* and other technology-oriented magazines would have noticed the increasing number of articles about recently launched space commercialization ventures. The goals of these new small-scale and privately owned companies resonated with the citizen-oriented approach to space exploration that groups like the L5 Society favored. They also comported with one of the Reagan administration's main goals for its "conservative space agenda": increasing commercial space activity.

The Reagan administration's promotion of greater space commercialization extended to individuals who wanted to build privately financed space vehicles and use them to put people and payloads into orbit. This signaled a shift for the pro-space movement as its initial goal of bootstrapping citizen settlements into space evolved to the pursuit of short-term political and economic aims. Dreams of building settlements in space were replaced by less utopian schemes such as putting satellites in orbit more cheaply than NASA could.

Like the biotech boom, the rising interest in space commercialization must be seen in the context of the pro-business and pro-technology laws and attitudes that emerged at the end of the Carter administration and continued full bore after Ronald Reagan took over the Oval Office. This entrepreneurial activity reflected a growing national embrace of materialism and financial success that marked the "greed is good" sentiments of the 1980s. Prosperity, in the newly emerging Reagan ideology, was patriotic, and optimism among technology enthusiasts ran high. "I'm going to be a billionaire," L5's Keith Henson boasted early on to one news reporter. "A lot of us are."[90] Even NASA got in on the act as it marketed its space shuttle fleet with a glossy brochure titled *We Deliver*.[91]

Encouraging space commercialization was one pillar of the Reagan administration's plan for outer space. Another was a greater focus on the militarization of space. The apotheosis of this policy was the Strategic Defense Initiative (SDI), unveiled by Reagan in March 1983. (SDI was more commonly and often derisively called "Star Wars," after the hugely popular movie by that name, because it called for high-tech defensive missiles, lasers, and other weaponry that would destroy incoming nuclear warheads before they could reach their targets.)Advocates of the plan believed that the advanced technologies developed for SDI held the possibility to render nuclear weapons "impotent and obsolete."[92] One might even think of SDI as a technological solution to the problem of arms control, a solution that scientists like Edward Teller offered to U.S. diplomats and politicians as a way to transcend the "limits" of negotiations, which were not making especially swift progress in reducing the stockpiles of U.S. and Soviet nuclear weapons.

Back in the pre-SDI world, years before scientists, policy makers, and the public considered the program's feasibility, the grassroots pro-space movement had begun debating whether the colonization of space should have a military component. In May 1976, for example, *L5 News* speculated about the possibilities of modifying future solar power satellites into military platforms and using their capacity to beam microwave energy back to earth as a weapon. Even among mainstream scientists, this wasn't such a far-fetched idea. One of the main Soviet objections to Star Wars was that, rather than playing defense, its systems might instead be converted into first-strike offensive weapons.[93]

Such discussions made some L5 members uncomfortable. One member asked that L5's newsletter "not stress this [space warfare] any further."[94] Keith Henson, speaking as a member of L5's board of directors, replied that the group would drop the issue only after "our under-sand respirators arrive." And the issue wouldn't stay buried. Another "long time space advocate and futurist" asked that *L5 News* at least present a mix of balanced and "socially conscientious" viewpoints to counter the more militaristic essays. The "betterment of the human race as a whole" was, as he saw it, L5's main goal. "Isn't that what we are seeking?" he asked. "If it isn't, damn you."[95]

The question of militarization eventually became the group's "most consistent source of controversy."[96] To be sure, some of the attention L5's writers gave to space-based weapons stemmed from the fascination with space battles that movies like *Star Wars* inspired.[97] Lasers, in the 1970s, were no more than a few decades old and still proved a reliable source of wonder.

But L5's increased attention to space industrialization, space commerce, and the militarization of space also reflects the group's gradual drift away from a counterculture-flavored enthusiasm for celestial communes where one could pursue alternative lifestyle choices. With the focus on such far-out communities fading, articles about space-based weapons, aerospace projects, and space industry came to dominate L5's newsletter. Over time, this shift amounted to tacit approval of Reagan's conservative space agenda.

This change of agenda further reflected not only a competition between different visions of the technological future but also between two competing means of achieving those imagined futures. On one hand was the quixotic idea of people leaving the planet by virtue of their own bootstrapping techniques. But a growing number of L5's members began to imagine instead that government funding and military activities in space could help open the space frontier. Analogies made to the nineteenth-century American frontier, in which forts, military expeditions, and government-funded railroad ventures presaged civilian settlements, were common. And, if NASA wouldn't fund solar power satellites and space stations, then maybe the Department of Defense would. As one L5 member wrote, "Even if I have to cut my hair to play space-marine for Uncle Sam to get out there, it's okay. The main idea is to get there. Take the High Ground!"[98] Many pro-space enthusiasts espoused a libertarian ethos reflected in their anti-*Limits* sentiments. Yet more than a few L5ers were also willing to have "Uncle Sam" and the military lead the way to space settlements, a view that helps explain L5's political shift rightward. To be sure, these views may have seemed somewhat schizophrenic to outsiders, but such a plurality of views was understandable given the pro-space movement's political diversity. As late as 1984, a congressman who spoke at a space development conference observed that L5's members were "five percent Democrat, five percent Republican, and ninety percent anarchist."[99]

In the new, militarized "space war" scenarios, O'Neill-style visions of peaceful space settlements were largely absent. Instead, space was framed, as had been the case after the Soviets' launch of *Sputnik 1*, as the "high ground," and gaining control of it became a matter of national survival. The conservative space movement even conscripted the title of O'Neill's book *The High Frontier*. In 1981, with underwriting from the Heritage Foundation and conservative donors, a retired general named Daniel O. Graham founded the High Frontier Project, which advocated for space-based weapons. Graham later claimed the name choice was accidental. O'Neill's personal papers are silent as to how he felt about having his own phrase co-opted. But, given O'Neill's initial interest in the peaceful humanization of space, his general leftward leanings, and the tenacity with which he protected his ideas, it is unlikely he approved.

When it started, the L5 Society was, one journalist recalled, a group of "Timothy Leary and *Whole Earth Catalog* people ... interested in social experimentation," who were "trying to make the world better."[100] But, when Carolyn and Keith Henson divorced in 1981, their split seemed to reflect the general divide among L5's members. Carolyn married a conservative defense analyst and began advocating space-based weapons and the conservative space agenda. Keith, ever more interested in cryonics, learned basic surgical techniques so he could assist in "de-animation" procedures and married Arel Lucas, an associate of Timothy Leary's.

The divisions among L5 members prompted some to wonder whether they had "gradually lowered our sights over the years." It was, one member wrote, "hard to keep up the drive for so long" when burdened with such a bold initial vision. "The dream's still alive with me," he said, "but it's getting rather weak."[101] The L5 Society did eventually disappear, but not with the mass disbanding in outer space its charter members had hoped for. Instead, in 1987, it merged with the much larger, wealthier, and corporate-oriented National Space Institute to form the National Space Society. Today, the group claims some twelve thousand members.

Gerard O'Neill never lost his enthusiasm for space exploration. But, by the time Reagan was running for president, O'Neill was no longer stumping for space colonies with the same vigor he displayed five years earlier. His own professional goals had shifted too. In 1979, after Congress held hearings about the military's satellite-based global navigation system—the government was considering making it available for civilian use—O'Neill submitted a patent application for a "satellite-based vehicle position determining system."[102] After his patent was approved, he relinquished his Princeton professorship to become a full-time technology entrepreneur.

Geostar, the company he founded in 1983, was based on developing a "radio-determination satellite system." O'Neill imagined that the company would sell location fixes, navigation aids, and tracking information in a manner similar to that of the military's classified global positioning system. Over time, O'Neill had plans that would also allow Geostar's subscribers to send and receive messages, including emergency requests for help, in addition to getting basic navigational information.[103]

As with all his projects, O'Neill thought big. Estimates for building the Geostar system, which would require at least two orbiting satellites, ran as high as $300 million. O'Neill set about raising money and recruiting partners for his venture. At one point, the roster of investors included the Nobel Prize–winning physicist Luis Alvarez, former chief NASA administrator Thomas Paine, and Bernard Oliver, who had directed research at the electronics giant Hewlett-Packard for years. Despite perennial cash shortages and opposition from various federal agencies, Geostar eventually conducted initial ground tests and raised millions of dollars to support further technology development.[104]

O'Neill's attentions also turned to broader, more mainstream issues of technology and national prestige. For example, his 1983 book *The Technology Edge* played on then-powerful anxieties that the United States was surrendering its technological preeminence to Japan (the original title was *Techno-Nationalism*).[105] O'Neill focused on several high-tech areas including robotics and "microengineering" and based his analysis on fact-finding trips to high-tech corporations in the United States, Japan, and Europe. However, *The Technology Edge* lacked *The High Frontier*'s originality and creativity. Instead, it fell into the expanding genre of technology-oriented books that fretted about American entrepreneurs' struggles to compete effectively with their Japanese and European counterparts.[106]

O'Neill's business ambitions were curtailed in 1985 when doctors diagnosed him with leukemia. Lengthy legal battles over Geostar's management followed, and O'Neill's company gradually collapsed as he struggled with chronic health issues. In 1992, O'Neill passed away in California. The obituary that his friend Freeman Dyson wrote praised O'Neill's visioneering, noting his ability to combine "far-reaching visions with practical work." Dyson characterized O'Neill as a freethinking scientist "unwilling to swim with the stream."[107] And though his visioneering for the humanization of space and Geostar had failed for "financial and political reasons," Dyson pointed out that none of O'Neill's ideas were discredited on their technical merits.

In 1997, O'Neill made national news one more time. On April 21, a specially modified airplane took off from Grand Canary Island. Strapped to its underside was a fifty-five-foot rocket to which engineers had bolted a canister laden with twenty-four small aluminum containers. One of the lipstick-size capsules was packed with seven grams of O'Neill's ashes. Other vials had remains of Timothy Leary, *Star Trek* creator Gene Roddenberry, German rocket engineer Krafft Ehricke, and Todd Hawley, a young man who had cofounded Students for the Exploration and Development of Space. "They look like little cocaine vials," one of the people watching the launch said, "which is kind of hysterical in Timothy's case."[108]

The plane climbed to 37,000 feet before releasing the rocket over the coast of Morocco. Ten minutes after its first stage ignited, the payload entered orbit, where O'Neill's ashes circled the earth every 96 minutes until they reentered the atmosphere five years later. By this time, the people inspired by his visioneering had turned to new technological frontiers. Merging ideas from biotechnology, computing, and microelectronics, nanotechnology beckoned.

CHAPTER 5

Could Small Be Beautiful?

IIIIIIIIIIIIIIIIIIIIIIIIIIIIIIIIIIIIIIIIIIIIIIIIIIIIIIIIIIIIIIIIIIIIIIIIIIIIIIIIIIIIIIIIIIIIIIIIIIIIIIIIIIIIIIIIIIIIIIIIIIIIIIIIIIIIIIIIII

How nicely microscopic units yield
In Units growing Visible, the World we wield!
  —John Updike, *Dance of the Solids*, 1969

Nanotechnology did not appear ex nihilo. Even if one could point to a single entity that encapsulates such a sprawling technological, scientific, and economic enterprise, nanotechnology is far too complex for one research discipline, let alone one person, to lay claim to it. But Eric Drexler and a cohort of like-minded technology enthusiasts did help create *something* new and exciting.

As the excitement of the space program waned, new technological frontiers were found not beyond our planet but with the manipulation of matter at the smallest scales. Instead of imagining a future that started with settlements floating in the inky vacuum of space, this new future derived from manipulations of the cell's interior machinery and the integrated circuit's crystalline architecture. Scientists and engineers combined their increasing abilities to engineer organisms and devices with an interactive and interventionist approach to modifying the physical world in new and unexpected ways.

Biotechnology and microelectronics provided a core foundation for activities that Drexler initially called "molecular engineering." Only later did these acquire the "nanotechnology" moniker. Drexler skillfully drew on existing concepts and technologies and merged them with his own ideas to synthesize a new and compelling vision of the technological future. The strategies Gerard O'Neill used to promote his ideas certainly afforded Drexler examples to draw on. The pro-space movement also provided Drexler with a community sympathetic to visioneering bold ideas for the technological future. The surge of new science- and technology-oriented venues such as *Omni* and new tools like the Internet helped ensure that ideas about nanotechnology could reach a much wider audience than would have been possible a decade earlier.

Investors and researchers recognized, of course, that microelectronics and biotechnology were complex ensembles of knowledge, tools, and techniques. Not rooted to any one thing or process, those fields of technology offered instead a general foundation for new ways of making things. In the 1970s, with traditional American manufacturing facing increased foreign competition, futurists and policy intellectuals across the political spectrum believed that emerging technologies like microelectronics and biotechnology would prove central to America's economy for decades to come. Consequently, when nanotechnology started to gain a public identity in the early 1980s, investors, scientists, and policy makers were already primed to see its potential.

But how did nanotechnology acquire visibility and begin to get attention from researchers, policy makers, and the public in the first place? One answer lies in the visioneering of people like Drexler who imagined a future reshaped by nanoscale engineering and then did research to help advance this vision. Microelectronics, computing, biotechnology, and even space exploration—each, in its own way, contributed something to visions for a future made possible by nanotechnology.

(Solar) Sailing toward Nanotechnology

When public interest in Gerard O'Neill's humanization of space ideas peaked in 1977, Drexler was just finishing his undergraduate degree at MIT. As Stewart Brand described him, he was already an "old Space Colony hand." A true believer in O'Neill's vision, Drexler had even once predicted "I probably won't die on this planet."[1] A photograph taken circa 1977 shows the young man at the "L5 society commune" in Tucson accompanied by a dairy goat, various garden vegetables, and a pile of scrap iron gathered for recycling.[2]

Drexler regularly wrote technical articles for L5's newsletter about asteroid mining and solar power satellites as well as more expansive pieces about the political and economic implications of space manufacturing and settlement. He also served on L5's board of directors for several years.

Drexler blended an MIT education with humanistic appeals to entertain new ideas rather than reject them outright. "To those who have looked at the night sky and not seen a black wall, blessings," he wrote Brand. "Life proceeds and spreads before the Sun." Like Brand and O'Neill, Drexler wanted to reconcile environmental and social concerns with ambitious technological visions. "Small may or may not be beautiful," he wrote, riffing on E. F. Schumacher, adding that "*good* is beautiful and varies with the situation."[3] A decade later, however, "small is beautiful" assumed an entirely new meaning for him.

In the 1980s, after Drexler became internationally recognized for his popularization and promotion of nanotechnology, journalists described him with classic stereotypes of an otherworldly science prodigy. Likening him to a "young Carl Sagan," one writer noted that his slow and deliberate speech was offset by "earnest intensity" and a "nervous affect." He was clad in "disheveled garb," observed another, and his "conspicuously expressive eyebrows" were set under a head of "frizzed hair cut like the young Einstein's." More comical were descriptions likening him to the "star of *Gilligan's Island*" and noting he was famous for wearing "zorries [a type of sandal] on his feet all winter" in Boston. Less common but more problematic were dramatic (or melodramatic) allusions casting Drexler in the role of ascetic prophet—"wan and exhausted," a man "startled by his own vision of the future," someone who had "just returned from a long trip . . . with a strong sense of mission in life . . . concerned that his ideas be taken seriously."[4] Pictures of Drexler in the mid-1970s, however, often just show a slender young man, sometimes goateed, with brown hair pulled into a ponytail and perhaps wearing a T-shirt sporting mathematical equations.

Born in 1955 near Oakland, California, Kim Eric Drexler grew up thinking, at times even worrying, about the technological future. A photo taken at the 1964 World's Fair in New York places nine-year-old Eric in front of the Unisphere, the massive model of the earth crafted from stainless steel that towered over the fair's lakes, lagoons, and reflecting pools. His parents were both science fiction fans who read to their young son about space exploration and astronomy. Drexler's mother was also a mathematician and science buff with engineering talents. She tackled home improvement jobs with aplomb and, during World War Two, helped test aircraft engines. Following his parents' divorce, Drexler moved with his mother to Denver. Her brother helped start a planetarium there, and his enthusiasm for astronomy left an impression on Eric. Eventually, mother and son settled in Oregon's Willamette Valley.[5]

Drexler's deep enthusiasm for science and technology continued into his teen years. A high school friend recalled how they rode their bikes for miles to take advantage of the library collections at Oregon State University in Corvallis. Drexler spent hours reading books or doing homemade science experiments like seeing, in anticipation of agriculture in space, whether plants could survive inside a refrigerator. Eric, according to his mother, could express an "eternal perspective," meaning that he tended to consider things in a global framework that peered far into the future.[6] He also had a "concern with large-scale issues" heightened by reading Rachel Carson's *Silent Spring*. He maintained a sense of humor, though. When other kids his age were wearing necklaces with peace signs, Drexler used masking tape to give himself a peace-symbol tan on his chest.

Drexler's encounter with *The Limits to Growth*'s warnings of near-inevitable societal and environmental collapse left him despondent. But, as it had for O'Neill, *Limits* served as a stimulant as well. "The book scared me," he said, "and helped crystallize my resolve to work toward a future outside the range of those scenarios."[7] To the young technology enthusiast, an "obvious way out of the box was to use space resources. It eliminated a fundamental premise of *Limits*, which is that of a well-defined, bounded volume of space and material."[8]

When he composed his application for admission to MIT, Drexler speculated that establishing human settlements and factories in space could be the "moral inverse of genocide," putting "mankind beyond accidental destruction or final collapse."[9] The same far-reaching perspective appeared later in Drexler's declarations of nanotechnology's perils and promises, claims some found ardent and others messianic.

Drexler's undergraduate major at MIT—interdisciplinary science—was originally designed to give students the opportunity to develop majors, with faculty guidance, that were "not provided for by regular departmental programs."[10] Given Drexler's wide range of interests, this choice fit him well. It encouraged a broad reading of the scientific and engineering literature, which, in turn, helped inspire the early expositions of what he first termed "molecular engineering." The path to nanotechnology did not appear suddenly, however. After he finished his undergraduate degree at MIT, Drexler stayed on to pursue a master's thesis in the Department of Aeronautics and Astronautics. For his research topic, the young engineer decided to design a "high performance solar sail system."[11]

A solar sail is an elegant idea. Light from the sun carries not just electromagnetic energy (perceived as the light we see, the heat we feel) but also physical momentum. This gentle but constant force, if captured by a sufficiently large sail, can propel a spacecraft until it is traveling hundreds of thousands of miles an hour. Solar sailing can only begin, however, once the spacecraft has been boosted by another means, such as a rocket, to a velocity sufficient to escape the earth's atmosphere and gravity, roughly 25,000 miles an hour. At that point its vast but gossamer-thin sail would unfurl and the craft would gently strain forward, gaining speed very slowly on its months- or even years-long journey of acceleration.

The concept of a solar sail is not only an elegant idea but also an old one. Russian space visionary Konstantin Tsiolkovskii described the basic idea well before 1930. It was first picked up in the United States by science fiction writers. Then, a few weeks after the Soviets launched *Sputnik 1* in October 1957, wunderkind physicist Richard L. Garwin added the rigor of numbers and equations.[12] Despite the solar sail's hypothetical potential for moving a spacecraft about the solar system, however, rockets remained the prevailing technology for space exploration during the Apollo era.

In mid-1976, as NASA recovered from its post-Apollo hangover, Bruce C. Murray, the new head of the Jet Propulsion Laboratory in Pasadena, proposed several new missions that he believed had "good technical content and popular appeal."[13] One idea Murray suggested was a rendezvous with Halley's Comet, scheduled to return to the inner solar system in 1986, with a craft powered by a solar sail. If successful, the propulsion method could perhaps be adapted for other NASA missions.[14]

The idea of solar-sailing a spaceship to meet Halley's Comet captured the public's interest.[15] As opposed to dirty, noisy rockets, with their obvious connections to ballistic missiles, solar sails appeared to be a "soft" technology akin to solar and wind energy. But, in the end, high inflation and budget deficits scuttled the entire mission, and no American craft flew to intercept Halley.[16] A small group of persistent NASA engineers, however, started the World Space Foundation to pursue their goal of building and testing a solar sail prototype using funds from private sources, and efforts to find a gentler alternative to rockets continued into the 1980s.[17]

This milieu—exciting and rife with possibilities—framed Drexler's visioneering for solar sails. He imagined the technology could prove essential for future space exploration efforts. Funded by a graduate fellowship from the National Science Foundation, he investigated what the best materials would be for a solar sail and designed a machine, compact enough to fit inside NASA's space shuttle, for fabricating ultrathin sheets of metal films in space. Drexler's visioneering also included a system to rig these pieces into a solar sail a few kilometers wide and a method for maneuvering the craft. To make the basic proof of concept more robust, he even fabricated some small pieces of metal films in an MIT lab.[18] Drexler dubbed the overall design concept "Lightsail," to make it clear that it would be propelled by photons, not the solar wind or some adaptation of solar power cells. As he joked, the moniker "sounds nice and my mother likes it."[19] Drexler thought the research he did at MIT was innovative enough to warrant filing for a patent.[20] And he promoted his ideas for solar sailing to L5 members as well as the wider public.[21]

Several years later, Drexler began using the term "exploratory engineering," an expression that well describes his approach to designing solar sails. Unlike conventional engineering, which aims to build specific things and looks to a relatively short-term horizon, exploratory engineering meant, for Drexler, "designing things that we can't yet build."[22] Although the concept shares some essential common ground with visioneering, the latter is even more ambitious in its action and goals. Visioneering certainly embraces a probing and exploratory approach to design as technically fluent experts speculate as to what *can* be built. Yet visioneering also involves the popularization of ideas, the construction of networks of supporters, and the cultivation of patrons. This combination of activities is what visioneers use to nudge society toward the expansive scenarios of the technological future they imagine.

Throughout his career, Drexler was careful to mark the boundary between lab-based scientific research and engineering. Engineering, he said, "aims to make things work" and is about "doing rather than discovering."[23] As he later explained, "You're not trying to learn more science. You're taking the textbook facts and saying, 'Provided this is true and given the methodology of engineering, what does this say about things that are possible?'" The history of space-related technologies exemplified for Drexler the power of exploratory engineering. People like Tsiolkovskii, Goddard, and von Braun "worked all sorts of hypotheticals in great detail," he said, "sometime before they had the hardware that could come close."[24]

Exploratory engineering, like visioneering, rested on a solid foundation of scientific facts and engineering principles. It was, Drexler said, accompanied by "rigorous criticism" to separate "the solid from the erroneous," which in turn sharply differentiated it from science fiction and futurists' vague speculations.[25] This proved a fine line to walk. "When a scientist indulges in speculation," the numerically scrupulous O'Neill said, "he throws away the experimental tools which give him his only claim to authority and expertise."[26] Visioneers may well be asked to produce tangible signs of progress toward the technologies central to their vision for the future. O'Neill's work on mass-driver technologies, along with the increasingly detailed NASA studies to which he contributed, helped to bolster the integrity of his visioneering. Years later, as we'll see, many mainstream researchers subjected Drexler's conceptualizations of nanotechnology to withering criticism. Disagreements over what counted as proof of concept and evidence of progress fostered opposition to his visioneering. However, before nanotechnology could be something worth fighting over, Drexler had to present his ideas to a wider audience.

Machines Make Other Machines, Repeat . . .

When Drexler first read *The Limits to Growth*, he concluded that escaping its premises would require a "profound change in the basis of our industrial civilization."[27] At first this meant looking to O'Neill's "high frontier." But, starting in late 1976, Drexler began to consider visioneering the technological future in a new way.

As Drexler later recalled in interviews with science writers, he began his anti-*Limits* journey with the unrevolutionary observation that "miniaturization is a big theme in technology."[28] Computers and their circuits, as Gordon Moore predicted, had been shrinking in size while increasing in performance for years. A logical extension of miniaturization would be to take advantage of the small size of organic molecules (a double helix of DNA, for example is less than fifteen nanometers in diameter, far smaller than any 1980s-era transistor) to build the elementary parts of a computer circuit. "Atoms and molecules are the ultimate building components," he remembered thinking. "I'd found engineer's heaven."[29]

One intellectual point of departure for Drexler was biologists' recent research on proteins. In both their structure and their chemistry, these macromolecules are extraordinarily complicated. Imagine a protein as a long necklace with dozens or even thousands of beads. Each bead is one of twenty different amino acids hooked to the next via a peptide bond, creating a long chain. The unique sequence of amino acids determines the shape and function of each protein. In reality, proteins fold into complex three-dimensional shapes as the bumps of some amino acids fit into the folds of others. Not only are proteins one of the basic building blocks of cells; their interconnected and coordinated actions also enable cells to carry out myriad functions. For example, enzymes are proteins that catalyze and control chemical reactions within individual cells.

Drexler raided MIT's library stacks for books and articles about molecular biology as he thought about how biomolecules might carry out a computer's basic logic functions. In magazines like *Scientific American*, it wasn't hard to find analogies that connected proteins and cellular organelles with electronics, computing, and factory machinery. In the biological realm, ribosomes inside cells make proteins that spontaneously assemble into the correct structures, which then form the basis for more complex forms. "So, I started asking myself, 'Well, what if we could do things like that?'" Drexler recalled. "You'd have a molecular technology that could be used to manufacture a wide range of products"[30]

Inchoate as they were, Drexler's initial thoughts blended and borrowed basic concepts from mechanical engineering, computing, and molecular biology. As he later confessed, "I didn't have any marvelous insights. I just had this image of self-assembling things."[31] Starting with the idea that biomolecules could, in theory, self-assemble to form a basic computing device, he imagined an even more general application. Perhaps these hypothetical biomolecular machines could manipulate molecules to build other things. "If you iterate this process, of using machines to build better machines, it's pretty clear that you could build machines that could build copies of themselves."[32] With this insight, Drexler added another component to his evolving mechano-computing-biological mélange: self-replication.

In Enlightenment-era Europe, the construction of lifelike automatons had provoked intense debate about the boundaries between living creatures and mechanical devices. For example, in 1738, French inventor Jacques de Vaucanson unveiled a mechanical "defecating duck" that ate and shat like a real *canard*.[33] If Vaucanson's famous duck was deemed ingenious, then "what shall we think of an engine of wood and metal which can," Edgar Allan Poe wrote a century later, "compute astronomical and navigation tables?"[34] Moreover, the idea of machines making copies of themselves already had a long history. In the mid-seventeenth century, when Queen Christina of Sweden was weighing Descartes's argument that animals were another form of mechanical automata, she allegedly asked the philosopher to explain how a clock might reproduce.[35]

It was only after World War Two, however, that serious scientific theorizing about how to make self-replicating machines emerged. Much of this work emanated from the abstract worlds of computer design and programming and merged concepts from microelectronics, computer science, and biology. Of course, organic entities like microbes and cells self-replicate. But, as scientists soon showed, inorganic and inanimate artifacts can do likewise. Computer viruses, programs that copy themselves in order to infect other computers, are a self-replicating example many of us are all too familiar with.

Hungarian-born polymath John von Neumann is the person most commonly associated with early mathematical theories of self-replication. In 1944, von Neumann learned of efforts underway at the University of Pennsylvania to build an "electronic numerical integrator and calculator" that would compute artillery firing tables for the military. Drawn to this project, von Neumann proposed a general "logic architecture" for electronic computers. His design for how a sequential stream of instructions would be fetched, decoded, and executed became the logical basis for almost all computer designs, from giant IBM mainframes to the laptop machine I'm writing on now. After the war, von Neumann began to think about the relation between more complex computing machines and biological systems. He had no interest in building such a device but found the underlying logical and mathematical concepts fascinating. Could these new "electronic brains," von Neumann wondered, be analogized to their human counterpart?[36] These were not idle speculations. "Artificial intelligence"—a topic that Drexler later studied—emerged in the mid-1950s as a vibrant new area of Cold War research.

As an exercise in logic, von Neumann proposed a plan for self-replication. He started by imagining a physical machine with a mechanical arm nestled amidst a surfeit of spare parts. Following a set of instructions, the machine could pick up a part, verify that it was the right one—incorrect pieces would be tossed back into the pile—and, after finding another correct part, connect the two. Once it built a physical duplicate of itself, the first machine would then pass its instructions to its "clone" before ordering the new machine to activate its own code, starting the cycle anew. Proposed several years before Watson and Crick deciphered the structure of DNA, von Neumann's schema, worked out via lengthy mathematical proofs and abstracted further to a geometrical grid of "cellular automata," resembled the mechanisms of genetic transcription and translation.[37]

It wasn't long before other researchers tested von Neumann's abstract concepts by building actual models. In London, British geneticist Lionel Penrose and his mathematician son Roger made a system of wooden blocks to demonstrate how "units that assemble replicas of themselves shed light on molecular reproduction."[38] A physicist at Brooklyn College took the idea even further. After scavenging parts from model train sets, Homer Jacobson built a mechanical self-replicating system using two kinds of cars that he designated "heads" and "tails." Initially arranged in random order, this "reproductive sequence device" would circulate around tracks he laid out. If a head and tail connected on a siding, they could signal to another head and another tail to connect on an adjacent track. Mediated by a complex system of batteries, switches, and photoelectric cells, the process would clack along until the supply of cars and sidings was exhausted. When Jacobson wrote up his findings, he noted the "degree of analogy between electromechanical and biochemical self-reproducing systems" that his model revealed.[39]

With basic principles of self-replicating machines established and rudimentary examples "reproducing" in a Brooklyn lab, a few scientists speculated further on potential applications. In 1956, Edward Moore, a mathematician at Bell Labs in New Jersey (and no relation to Gordon Moore), proposed a hypothetical machine that could extract raw materials from its environment and then use solar energy to build a copy of itself. Perhaps such an "artificial plant" might even be designed "to make a product which was not useful" to the machine itself such as fresh water. Reflecting the bold ambitions computer researchers had in the 1950s, Moore, who went on to do research on automata theory at the University of Wisconsin, predicted such machines would be "more easily attainable than human flight to other planets in a spaceship."[40]

Other scientists offered similar conjectures about the potential use of specially engineered apparatuses that could both self-replicate and make useful stuff. Physicist Theodore Taylor, renowned for his innovative designs of nuclear weapons, called such hypothetical devices "Santa Claus machines" and discussed them with Freeman Dyson when the two men worked together on Project Orion, the nuclear-propelled spaceship.[41] Self-replicating machines, hypothetically speaking, could also pose dangers. If not controlled, such machines might replicate exponentially and quickly "fill up the oceans and continents," an idea that presaged concerns about genetically modified biomolecules in the 1970s or, later, fears of nanoscale machines reproducing unchecked.[42]

In the late 1970s, the possibility of self-replicating machines caught the attention of Robert A. Frosch, NASA's new head administrator. Speaking at the Commonwealth Club in San Francisco, Frosch suggested that self-replicating machines might be vital to the future of planetary exploration as well as our terrestrial economy. Frosch started with the assumption that securing access to energy and materials "out in the solar system" was one way to overcome the *Limits to Growth* thesis but that conventional approaches to accessing space resources were too expensive. As a solution, he proposed "machines which can construct generation after generation of machines . . . in a pseudo-biological way." The result could be a "productive machine economy in an extraterrestrial place," setting the stage for an O'Neill-style expansion into the solar system.[43] With *Omni* reporting on it, news about Frosch's speech percolated throughout the pro-space community. In June 1980, NASA even brought together two dozen scientists and science fiction writers to kick off a study at the University of Santa Clara in Silicon Valley on "advanced automation for space mis-

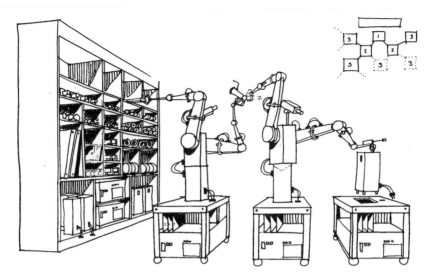

Figure 5.1 Illustration from the cover of *Advanced Automation for Space Missions*, a 1982 NASA report. (Image courtesy of NASA.)

sions." The project produced a report whose cover featured a von Neumann–like image of industrial-looking machines building more and smaller copies of themselves.[44]

Not surprisingly, the idea of machines autonomously making other machines intrigued science fiction writers. In 1964, for example, Polish author Stanisław Lem published a novel called *The Invincible*. Its story line unfolds as a spaceship's crew lands on a seemingly uninhabited planet and instead discovers a species of self-replicating micromachines that can self-assemble into intelligent complex swarms.[45] Perhaps images of self-replicating machines, proliferating autonomously and uncontrollably, reflected fears of the world's ever-larger nuclear arsenals or the unconstrained growth of human populations. To be sure, imagery of unrestrained self-replicating machines appeared in dozens of popular accounts describing Drexler's nano visioneering.

As Drexler mulled all these ideas over he also discussed them at great length with Christine L. Peterson, his close companion and intellectual partner. Peterson was born in Buffalo just after the first Sputnik launch. Her father, a mechanical engineer, did a lot of hands-on tinkering at home, and her high school chemistry teacher made a big impression as well. Besides math and chemistry, Peterson was especially interested in ecology and environmental issues. Like O'Neill, she saw the expansion of human activity into space as a way to relieve environmental pressures on earth. Peterson met Drexler at MIT, and they assumed active roles in their local L5 chapter. Peterson eventually became the L5 Society's national chapter coordinator and secretary. When she finished her undergraduate degree in 1979, the national economy was in a poor state, but she landed a job in the Boston area as a semiconductor engineer. They wed two years later and were married for more than twenty years.[46]

While Peterson helped design microwave diodes, Drexler maintained his graduate studies, thinking perhaps he would get a doctorate in aeronautics from MIT. Meanwhile, he continued to read widely about molecular biology, electronics, and computing and the technological spaces in which they might overlap. In 1979, Drexler happened across the November edition of *Physics Today*, which was a special issue devoted to "microscience." According to Lewis Branscomb, IBM's chief scientist, this term referred to the "new and fast developing interdisciplinary science" that provided the foundation for a "worldwide growth industry of extraordinary technical vitality."[47] The articles in *Physics Today* described engineers' ability to fabricate new devices on scales of less than a micron (a millionth of a meter or a thousand times bigger than a nanometer).

But what really caught Drexler's attention was that *Physics Today* started its issue with an eye not toward the technological future but with a reference to the past—namely, a reprint of Richard Feynman's 1959 "There's Plenty of Room at the Bottom" speech. Its resurfacing in the physics magazine initiated the essay's climb to prominence and a high citation count.[48] Drexler eagerly read Feynman's series of "what ifs," which blended the mechanical, the biological, and the industrial. Feynman spoke of a "hundred tiny hands" that could build a "billion tiny factories, models of each other, which are manufacturing simultaneously, drilling holes, stamping parts, and so on."[49]

Feynman's idea, in turn, for these "tiny hands" drew on conversations he had with his good friend Albert R. Hibbs, a rocket engineer at the Jet Propulsion Laboratory.[50] Hibbs had recently read a novella called *Waldo*, written in 1942 by Robert Heinlein. The protagonist in Heinlein's story, a genius engineer named Waldo Jones, devised a series of mechanical hands that duplicated the motions of their disabled builder's hands. These devices resembled the "remote manipulators" that real-world lab technicians used to work with hazardous or radioactive materials. Hibbs talked with Feynman about robots that could make smaller versions of themselves or perhaps even a miniature "mechanical surgeon" that a patient could swallow, a common trope that reappeared frequently in journalists' accounts of nanotechnology.

Future computing and information storage possibilities were also central to Feynman's vision. He hinted at how all the books in the world could be reduced to the space of a few dozen conventional-size pages and how circuits and computers might be built using quantum mechanical phenomena. Discussions Feynman had with MIT's Philip Morrison (the physicist who first put Drexler in touch with O'Neill) reflected au courant ideas about cybernetics and information theory. Meanwhile, Feynman alluded to techniques such as electron beam etching and vapor deposition that the microelectronics industry used to "write" increasingly smaller circuit patterns.

Finally, Feynman was "inspired by biological phenomena." Six years before he wrote his after-dinner talk, Watson and Crick had revealed the basic structure of DNA (a goal also hotly pursued by Linus Pauling, one of Feynman's Caltech colleagues). So it was natural for Feynman to note that biomolecules like DNA could store "enormous amounts of information . . . in an exceedingly small space." Eventually, the ultimate engineering would happen, Feynman said, "in the great future" when people would be able to "arrange atoms, the very atoms, all the way down!"[51] Moreover, Feynman vouched that the atomic engineering he imagined would violate no laws of science but instead presented challenges for his colleagues and students to overcome.

The Caltech physicist wasn't the only one thinking along these lines. At MIT, émigré physicist Arthur von Hippel predicted that the next technological revolution would occur when engineers possessed greater control over the material world. Especially interested in developing new materials for electronic applications, von Hippel began to advocate, a full three years before Feynman's speech, "the building of materials and devices to order." What von Hippel called "molecular engineering" meant the ability to deliberately fashion the materials one wanted "from atoms and molecules." This stood in contrast to the older empirical approaches to materials science, techniques von Hippel derided as "bulldozer tactics."[52] Von Hippel had lab results to support his view. In the latter years of World War Two, he discovered that crystals of barium titanate had special electrical properties that might allow them to be used as switches and memory elements for digital computers.[53]

It is possible Feynman was aware of von Hippel's vision of molecular engineering, as the latter's ideas were published in the flagship journal *Science*. To be sure, von Hippel's goals—the development of better materials for mechanical and electronic applications—were more circumscribed than Feynman's sci-fi-inflected descriptions of futuristic devices. However, both scientists envisioned a day when physics would eventually provide the means, as Feynman said, to "put the atoms down where the chemist says."[54] "The answer is therefore not any longer what we can do," von Hippel predicted, "but what we want to do."[55] What was missing in the 1950s, however, as both men acknowledged, was a toolbox of instruments and techniques that would enable the manipulation of matter at this fine scale.

Intrigued by Feynman's essay, Drexler turned to physicist Arthur R. Kantrowitz for support and critiques. Before taking a professorship at Dartmouth College, Kantrowitz ran a lab that researched, among other things, heat shields for missile nose cones, high-powered lasers, and technologies for artificial hearts.[56] In the late 1960s, concerned about political influence on scientific and technological decisions, Kantrowitz became a well-known advocate for what reporters dubbed a "science court." Its goal was to "separate the scientific from the political and moral components" of controversies such as nuclear power or recombinant DNA and provide a neutral basis to make "the great decisions" that determine "directions of progress."[57] An explanation of how "technical facts" might be cleanly cleaved from "values" never fully satisfied critics, but, given the public's ambivalence toward technology, Kantrowitz's proposal drew endorsements from elite scientists as well as officials in President Ford's administration.[58]

Drexler met Kantrowitz around 1975, and they both attended O'Neill's space settlement meeting that year in Princeton. Kantrowitz also encouraged Drexler's reading of works by Karl Popper, the famous philosopher of science. Kantrowitz's plan for a science court reflected Popperesque explanations of how scientists produced and verified new knowledge, abstract views that often ignored the context in which these activities occurred. In time, the elder physicist became a mentor of sorts to Drexler, who appreciated Kantrowitz's opinion that science worked best when separated from social and political influences. This overidealized and abstracted image of how scientific communities functioned caused considerable problems for Drexler years later when his exploratory engineering was attacked by some mainstream scientists.[59]

Throughout 1980, Drexler prepared a draft manuscript that he intended to submit to the *Proceedings of the National Academy of Sciences*, a journal published by the nation's foremost honorary society for scientists. Submissions to *PNAS* by people who aren't academy members are "communicated" to the journal by a member, a task Kantrowitz agreed to perform. Along the way, Drexler solicited scientific feedback on his draft from MIT faculty such as Jerome Y. Lettvin (a well-known cognitive scientist and electrical engineer) and Christopher T. Walsh (a rising star in biochemistry). Philip Morrison, meanwhile, encouraged Drexler to downplay "implications that smelled strongly of military usefulness."[60]

Drexler titled his article "Molecular Engineering" with the subtitle "An Approach to the Development of General Capabilities for Molecular Manipulation." The word "nanotechnology" appeared nowhere in the manuscript. In fact, it would still be a few more years at least before Drexler began using the *nano* word in talks and papers. The title he chose, though, was identical to one von Hippel had used in 1956. As Drexler later explained, he understood that von Hippel had written about "semiconductor technologies, not molecules as generally understood" and concluded that their respective approaches were "essentially unrelated."[61]

PNAS published Drexler's paper in September 1981.[62] Unlike other articles that appeared in the same issue, it contained no equations or formulas. Instead, Drexler offered his readers "close analogies between the proposed steps and past developments in nature and technology" rather than "mathematical proof."[63] The "general capabilities" his title alluded to were to be considered akin to early work like von Neumann's on the "theoretical capabilities of computers," which did not go so far as to predict "practical embodiments" such as actual hardware. Drexler likewise claimed his goal for "molecular technology" was not understanding all the complexities of natural phenomena but, instead, taking an engineer's good-enough approach to generate just enough knowledge "to produce useful systems."[64]

Although he acknowledged Feynman's 1959 talk (thereby enrolling the famous physicist's name), Drexler drew little primary inspiration from physics or the microelectronics industry. Instead, molecular biology gave Drexler his main font of ideas. As he often explained, nature itself offered the necessary proof of concept. In biology, "there are systems of molecular devices (cells) in which the collection of devices, working together, can make every kind of component. . . . The system as a whole can do this and then reorganize into two systems of the same competence."[65] Biology, in other words, was based on small, self-contained, self-replicating systems. Given the advances researchers were already making in what Drexler called "biochemical microtechnology," it followed that recombinant DNA and protein-engineering techniques would eventually allow the modification of cellular "ribosomal machinery" to produce "novel structures, which can serve as components of larger molecular systems."[66] This view combined the tools of the molecular biologist with an engineer's approach to building things, albeit at the molecular scale.

Despite the biological inspiration, Drexler's view of cellular structure and function remained highly mechanical. Motors and driveshafts, he noted, exist to move things in our everyday world while bacterial flagella and shape-changing proteins perform similar functions at the cellular level. Employing analogs between biology and machines had a long history, of course. Decades earlier,

Russian physiologist Ivan Pavlov used factory comparisons to understand digestive processes, while metaphors likening biological processes to the activities of machines (for example, women giving birth as "laborers" with their bodies as "machines" and doctors as "mechanics" to repair them) had an even older lineage.[67] After World War Two, molecular biologists spoke of genetic "codes" and cellular "factories."

Similar reasoning-by-analogy led Drexler to conclude that "mechanical systems can be constructed on a molecular scale." Such devices could, in theory, "move molecular objects" and "position them with atomic precision." The laws of nature, however, necessitate that artificial protein-based machines can exist only at moderate temperatures and in watery environments. To overcome this problem, Drexler suggested that the future protein machines could eventually build a "second generation of molecular machinery" out of more durable and adaptable materials such as diamond.[68]

But to what end? The applications Drexler suggested revealed how computing, electronics, and molecular biology coalesced in his thinking. One use Drexler proposed for "molecular technology" was storing and processing information. This recalled earlier attempts to build mechanical (as opposed to electrical) computers. For example, Charles Babbage had proposed a mechanical "difference engine" as early as 1822, and at MIT a century later Vannevar Bush had constructed a mechanical "differential analyzer." Drexler thought molecular-scale machinery could perform likewise.

Drexler's proposed biological applications were even more speculative. Given that the devices he imagined were minuscule, it was possible they might be able to repair cellular or other biological materials, for instance. Drexler concluded his exercise in exploratory engineering with the supposition—perhaps suggested by SETI proponent Philip Morrison—that the implications of molecular technology were important to "the probable behavior (and likelihood) of extraterrestrial technological civilizations."[69] Therefore, he advised researchers to consider the "opportunities and dangers" that the molecular technology might pose.

Getting a final draft of the manuscript ready had been an arduous task, but it compelled Drexler to tighten his argument.[70] The referee reports he received—one from Freeman Dyson and another from Philip Morrison—before *PNAS* would accept his article raised criticisms that scientists would continue to voice after Drexler's ideas for nanotechnology acquired a wider audience. Both physicists strongly recommended publication. However, as had been the case with O'Neill's original space colonization manuscript, they were less sanguine when it came to practicability. Dyson and Morrison, for example, critiqued Drexler's lack of empirical evidence. As Dyson put it, Drexler didn't "discuss the many questions of detail which much be addressed before one can judge the feasibility of the author's proposals."[71] Dyson also wondered why Drexler focused on designing new protein-based machines de novo rather than using genetic engineering to modify existing organisms. As Dyson expressed it, "in molecular engineering, as in macroscopic engineering, the horse should come before the steam engine."

But, as Morrison noted, Drexler's *PNAS* paper did successfully present an "imaginative" qualitative argument about a "novel domain of engineering not before practiced," one that might excite molecular biologists, protein chemists, and computer scientists.[72] Drexler was certainly aware that what he proposed spilled over into many scientific disciplines. The National Academy's rules required Kantrowitz to certify any article he communicated to be of "exceptional importance" and in the "top 10th percentile of its field." As Drexler wryly noted to his mentor, "If you can determine what field this paper is in, please tell me."[73]

Bold and speculative? Absolutely. Vague on details and a specific path to realization? Most certainly. Drexler's article, in the same spirit as Feynman's, proposed a host of things people could do as advances in microelectronics, computing, and biology converged "at the bottom." But, lacking detailed designs, it wasn't yet visioneering as we have defined the term.

Drexler's article resembled Feynman's in another way. After it appeared, it sank into the scientific literature with barely a ripple. Fewer than a baker's dozen of articles cited it before 1988. But, by this point, technology enthusiasts, former pro-space advocates, and science writers had transformed nanotechnology into the next big thing. It was a start.

Tomorrow's Tools

After Drexler published his first paper on molecular engineering in 1981, he entertained different career options. One was to complete his doctorate in aerospace engineering and perhaps join one of the new companies that were trying to get private spaceflight off the ground in the early 1980s. Another option was to shift priorities at MIT and start a program of laboratory research, perhaps in chemistry or materials science, that would produce some experimental evidence for the bold concepts he proposed. But instead Drexler opted to try to raise awareness and interest in the possibilities of "molecular technology" among other researchers and the broader public.

He began with the MIT community, giving a series of talks on a broad range of futuristic technologies including the ability to "build devices to complex atomic specifications."[74] One important contact Drexler had was the polymath Marvin Minsky. After receiving his doctorate in mathematics from Princeton in 1954, Minsky was one of the pioneers in artificial intelligence (AI). AI was a topic that interested Drexler, and while he attended MIT research in the field underwent a boom period, much of it financed by the U.S. military.[75] A good deal of this was carried out at the MIT lab Minsky had founded. Minsky became another important mentor for Drexler, and the elder scientist frequently supported his exploratory ideas.

With Minsky's encouragement, *Smithsonian* magazine invited Drexler to present his ideas to a general audience. Some two million people read *Smithsonian* every month, so Drexler's piece was guaranteed a wide readership. His article on molecular engineering appeared in November 1982. Throughout the piece, Drexler's choice of verb tense suggested a future in which a new technology that fused concepts from microelectronics and biology with self-replication and computer control was likely to exist: "advanced molecular machines *will* let engineers build molecular structures," and "engineers *will* build robot arms to handle individual molecules."[76] This rhetorical technique—what writers call prolepsis—appeared often in Drexler's writings as well as journalists' coverage of his ideas. It eventually became a noticeable trait that some scientists took issue with.

Smithsonian commissioned veteran illustrator John Huehnergarth to provide colorful images to go with Drexler's article. While the article gave readers a sense of what was possible, the pictures, in effect, said: This is what the nanotechnology of the future will look like. The illustrations and their captions were somewhat irreverent toward traditional chemistry—a scientist in an obligatory white lab coat was "blindly stirring" materials in a flask, suggesting a haphazard process compared to the required precision of Drexler's molecular technologies. The images also presaged those that became de rigueur in journalistic treatments of nanotechnology. For instance, Huehnergarth drew a "cell-repair machine" attacking a viral intruder. There was also an image of a "general assembly machine" reading a computer tape while its robot fingers directed the placement of molecules and atoms. These were not the elusive entities described by the complex workings of biochemistry or quantum mechanics but anthropomorphic machines whose "realness" suggested an ease of building with molecular Tinkertoys. Although Drexler himself may have objected to such simplistic, perhaps even distorted representations of his ideas, *Smithsonian* offered a wide audience to whom he could promote his view of the molecularly engineered future.

When the *Smithsonian* piece appeared, Drexler was spending long hours burrowed in MIT's libraries or discussing scientific and engineering concepts with colleagues. Biotechnology, including the nascent field of protein engineering, especially got his attention. For instance, a team of chemists led by William F. DeGrado designed and made peptides (chains of amino acids structurally similar to proteins but much shorter in length) with properties similar to those that occurred naturally in a particular bee venom.[77] A few years later, Carl Pabo, a biophysicist at the Johns Hopkins School of Medicine, described techniques for the "rational modification" of proteins and peptides so that they might take on desired structures and folding arrangements.[78]

Kevin Ulmer was another biotechnologist whose work inspired Drexler. Ulmer had grown up a few miles from Bell Labs, where the transistor and the laser had been developed. Just a few years

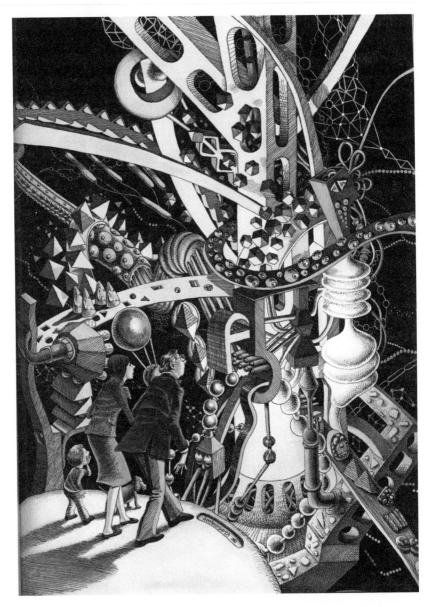

Figure 5.2 A fanciful interpretation, by artist John Huehnergarth, of Eric Drexler's vision for molecular manufacturing. As a cartoon family looks on with wonder, assembly lines bring "protein parts" together to be assembled "atom by atom" into a "giant machine." From the November 1982 issue of *Smithsonian*. (Image courtesy of the artist.)

older than Drexler, he received his doctorate in 1978 from MIT for his research on repairing *E. coli*'s DNA structure after it had been irradiated. A year later, he accepted a senior research position as "director of exploratory research" at Genex, a biotech start-up in suburban Maryland that grew rapidly when corporations and investors poured capital into the emerging biotech sector. Many of Genex's scientific staff focused on making amino acids for industrial use, a profitable but prosaic activity.[79] Ulmer had more ambitious goals. One was the potential "engineering of materials outside the range found in nature's catalogue."[80]

Ulmer first learned of Drexler's hypothetical protein machines when *Smithsonian* asked him to review a draft of Drexler's article. He saw that Drexler's ideas for integrating biology and microelectronics resonated with ones he and other researchers were exploring. In the early 1980s, there was discussion, for instance, about incorporating organic molecules into computer chips, perhaps even replacing silicon with protein.[81]

The idea of "molecular electronics" goes back to the 1950s, when engineers at Westinghouse and General Electric were trying to leapfrog competitors with bold new technological concepts.[82] Despite healthy Air Force support, these efforts collapsed in the mid-1960s as silicon-based circuits assumed industrial hegemony. Molecular electronics reappeared in the 1970s as companies like IBM contemplated building circuits using organic molecules in place of silicon transistors. This interest followed the discovery in the 1970s of organic polymers that could conduct electricity, work that became a hot new area of chemistry.[83]

Spurred by these scientific results, a chemist at the Naval Research Laboratory named Forrest Carter began to attract a small but enthusiastic group of researchers to his vision for molecular electronics. Trained at Caltech, Carter counted Linus Pauling and Richard Feynman among his mentors. Carter was one of the few scientists who had read Feynman's "Plenty of Room" speech before its rediscovery. A charismatic personality and fondness for exotic hobbies like platform diving helped Carter attract protégés and estrange supervisors.[84] Because (or in spite) of Carter's advocacy, the navy, the National Science Foundation, and industry funded work on "moletronics" for years. Eager to promote molecular electronics to reporters, investors, and other researchers, Carter organized several conferences on molecular electronics that both Drexler and Ulmer attended. In the early 1980s, journalists looking to connect the revolutions underway in biotech and microelectronics often presented concepts for "biochips" as the next big step in computing.[85] Before his death in 1987, Carter's visionary ideas received recognition as well as opprobrium from other scientists while his navy overseers began to demand that he produce more lab results and less blue-sky speculation.

Although some electrical engineers expressed skepticism that the silicon throne could be easily overturned, a few scientists imagined eventually making electrical circuits using specially designed and engineered protein molecules. "We can now view protein somewhat the way we do polyethylene, stainless steel, and silicon," Ulmer explained at a seminar on biotechnology. "It is a material we can shape and build with."[86]

Ulmer's optimism was enhanced by advances in "site-directed mutagenesis." Developed in the late 1970s by scientists like future Nobel Prize–winner Michael Smith, this technique was a logical extension of decades worth of research aimed at understanding the relationship between protein structure and function. As Ulmer described in *Science*, protein engineering could be done by starting with molecules whose structures were already established. After slightly modifying a protein's sequence of amino acids, researchers could then observe how it folded and what new reaction sites were exposed.[87]

Despite molecular biologists' ever-improving tools for studying proteins and sequencing DNA, correlating a protein's structure to what it did was an expensive and laborious process that typically entailed a cut-and-try approach. But, in the heat of the biotech boom, protein engineering was something several companies were willing to invest in. Eventually, these research efforts could serve as a stepping-stones, Ulmer said, toward a "general capability for molecular engineering" that might allow people "to structure matter atom by atom."[88] The research of scientists like Ulmer, Pabo, and DeGrado helped convince Drexler that his ideas about protein engineering and design were grounded in benchtop science that was already underway. Meanwhile the expansive view proffered by Carter suggested the kinds of applications and devices that molecular-scale engineering might produce.

By the end of 1983, Drexler finished a draft—provisionally titled "The Future by Design" —of what ultimately became his strong-selling 1986 book *Engines of Creation*. In it, Drexler outlined how continued advances in microelectronics and biotechnology would inevitably lead to atomically precise "control over the structure of matter" (the word *nanotechnology* still remained unused at this date).[89] Basically, this meant engineered "protein machines" that would build second-generation "machines of tougher stuff." These, in turn, would "assemble large molecules from small molecules under the direction of a molecular [computer] tape." Drexler called the hypothetical machines that could rearrange molecules into the desired patterns "assemblers," a term also used by computer engineers for instructions that convert a programmer's words into language that a computer can understand. He also articulated an "assembly principle" that stated, "molecular machines of the second generation will be able to build virtually anything that can be designed." To him, the potential power of these "assemblers" represented a distinct break with the technological past and presented a host of opportunities, challenges, and even dangers.

Compared with O'Neill's detailed estimations, calculations, and extrapolations, Drexler's ideas still lacked many specifics. Drexler suggested no experiments that could verify his concepts but instead described general ways in which assembler technologies might be achieved and the reasons why—citing Feynman, for instance—molecular technologies were indeed possible. Throughout his book draft, Drexler linked ideas for assemblers and protein machines to established practices in molecular biology and microelectronics. But he gave much more attention to all the myriad ways that assembler technologies of the future would revolutionize everything from space travel to computing to medical research.

Influences from outside the arenas of science and technology can also be seen in Drexler's protobook. Despite being more than a decade old and widely refuted by economists and other scholars, the pessimistic warnings of *The Limits to Growth* still offered a prominent catalyst. In fact, as had been the case for O'Neill, Drexler *needed* the alarming vision *Limits* presented. Although public debate about the *Limits* thesis had faded, that book provided a necessary and invaluable foil against which to present a different vision of the technological future.

Referring to the "doom book industry," Drexler took future-looking specialists like Paul Ehrlich to task for their "grey, grim, and ridiculous" predictions.[90] Drexler was especially critical of doomsayers like economist-activist Jeremy Rifkin who, he argued, misused scientific concepts like entropy to buttress arguments that the resources available to people were finite. Such Malthusians, Drexler argued, mistakenly viewed the earth as an isolated system and ignored the possibilities of tapping extraplanetary resources, vastly expanding the use of solar power, and new variables such as molecular technology or genetic engineering. Drexler instead favored "cornucopians" like Julian Simon—an economist (and *Limits* gadfly) famous for winning a 1980 bet with Ehrlich that prices for key natural resources would fall despite population growth—as well as the promise of accelerating technological change.

Another echo from the 1970s, this from Stewart Brand's *Whole Earth Catalog*, also infiltrated Drexler's writing. To support his assertion that "technology evolves" via invention and innovation, Drexler referred to the work of industrial designer and writer James ("J") Baldwin. Baldwin was a leader among the "thing-makers, tool freaks, and prototypers" who read *Whole Earth*.[91] For years, Baldwin had modified, lent, tested, and improved "one highly evolved toolbox" that he kept inside his workshop-on-wheels (a 1958 Chevy van). Over time, the collection, as Baldwin described it, transformed into a "thing-making system" rather than just "a pile of hardware."[92] Such a general "thing-making system" was exactly what Drexler envisioned his assemblers becoming. But, unlike the hammers, vises, and drills in Baldwin's mobile tool-shed, "the tools of tomorrow" did not "hang on the wall and beg to be used." For the moment, they still existed "in the mind and in the potential implicit in natural law."[93]

The philosophical and political influences on Drexler's thinking are clear in his draft. References to famous scientists and philosophers—Karl Popper, David Hume, Feynman, Robert Goddard, von Neumann, Benjamin Franklin—abound. But devoid of contingency or context, their appearance suggests a view of history in which scientific discoveries and technological developments can be planned and managed. One also finds shades of Kantrowitz's "science court" and Popper's arid view of the scientific method rather than the messy, politicized, and contingent nature of real-world research.

Since his undergraduate days at MIT, Drexler had become increasingly interested in certain classically liberal ideas. He especially liked those expressed by the Austrian economist Friedrich Hayek, famous for his defense of free markets, limited government power, and individual liberty. Drexler speculated that molecular technology might bring political benefits by uniting polarized groups such as environmentalists and industrialists. As Drexler saw it, assembler technology, like space colonization, was compatible with environmentalists' goal of cleaner manufacturing processes. Moreover, abundances that new technologies made available could defuse tensions between libertarians and socialists. To make his point, Drexler approvingly cited Robert Nozick, a Harvard political philosopher who wrote the 1974 libertarian manifesto *Anarchy, State, and Utopia*. Nozick argued for the primacy of individual rights and a minimal state, sufficient to protect against violence and enforce contracts. All other tasks—from education to space exploration—would best be done by private institutions operating in a free market. Such ideas had already found favor with many in the pro-space crowd; they would later reappear among the nanotechnology community that coalesced around Drexler and his ideas after 1986.

Test Tubes versus Microscopes

As Drexler wrote and rewrote what eventually became *Engines of Creation*, he sent drafts to colleagues at MIT and other technology enthusiasts. This included friends in the pro-space movement. For example, Keith Henson, a cofounder of the L5 Society, had a chance to read an early version. Henson was especially enthused by the potential medical applications of Drexler's molecular technology, and he offered to help publicize the ideas more widely via his friendship with Timothy Leary. Drexler demurred—perhaps recalling the questionable benefit of Leary's enthusiasm for space colonies—saying he believed his work needed a more "credible introduction" than the former LSD guru might provide.[94]

Conrad W. Schneiker was another L5er who offered Drexler generous suggestions and editorial critiques. Born a few years before Drexler and raised in Tucson, Schneiker was interested in all aspects of science and engineering from a young age. During the 1970s, he gradually worked toward an undergraduate degree in engineering mathematics from the University of Arizona, occasionally taking time off to work for electronics companies in California. He joined the L5 Society soon after it was founded and regularly contributed to its newsletter. Unlike many other L5ers, though, Schneiker preferred to focus more on science facts. "I was so enamored with science," he recalled, "Science fiction seemed to me to be all soap operas."[95]

Schneiker, like Drexler, voraciously read *Scientific American*, *Science*, and other more specialized journals. He spent long hours browsing in university libraries and bookshops while taking additional university courses. Schneiker also read an early version of Drexler's book. Besides offering friendly critiques of Drexler's prose and ideas, Schneiker shared citations for articles and books he thought might be relevant to his colleague. He especially encouraged Drexler to "give Feynman's approach the attention it deserves."[96] Schneiker had known about Feynman's ideas ever since he came across a 1961 book on miniaturization in which the physicist's paper was reprinted. To him, Feynman's "Plenty of Room" was an essential text for anyone thinking about "building small."

However, Schneiker began to imagine a different conceptual path for manipulating matter at the finest scales. Whereas Drexler's vision was firmly rooted in future developments that might take place in the biological realm, Schneiker drew much of his inspiration from established techniques in microelectronics and instrumentation.

For instance, Schneiker avidly read technical papers by Kenneth Shoulders, a largely self-taught researcher who worked in von Hippel's lab at MIT before moving to the Stanford Research Institute. In the 1950s, Shoulders had imagined fabricating new kinds of electronic circuits by modifying an electron microscope so it could "write" ultrafine features as small as a 100 nanometers.[97] For years, Shoulders pursued the goal of using electron microscopes as tools for *making* rather than just seeing extremely small objects, a line of research Feynman alluded to in his "Plenty of Room" talk.

But what especially excited Schneiker was the recent invention of a new kind of electron microscope. In 1979, Heinrich Rohrer and Gerd Binnig, two scientists at an IBM lab in Zurich, began work on what became known as the scanning tunneling microscope (STM).[98] Rather than the lenses and mirrors a traditional optical microscope uses to produce an image, the new microscope used a sharp tip to probe the surface of a metal or semiconductor sample. By creating a voltage difference between the probe tip and the sample and then bringing the tip very close to the sample, some electrons could be made to "tunnel" between the two. (Tunneling is a quantum mechanical phenomenon in which particles like electrons can pass through a barrier that would be insurmountable in classical physics.) If Binnig and Rohrer moved their probe tip back and forth over the sample's surface, they could measure the changing strength of the tunneling current. Then, by keeping the current constant and continuing to scan with the probe tip, they could capture and convert the electrical signal to produce an atomic-scale image of a sample's surface.

After Binnig and Rohrer published their results in 1982, a flood of publications about the new instrument's capabilities appeared in specialist journals.[99] Meanwhile, continued improvements provided the basis for hundreds of patents as entrepreneurial researchers commercialized the STM. Within a few short years, the STM and its variants became ubiquitous instruments in labs and factories around the world.[100]

Schneiker first learned about the STM around 1983 while he was working in the Los Angeles area for a small software company. He was fascinated with the instrument's potential for realizing Feynman's broad vision of not just seeing individual atoms—scientists had been able to do this in various ways since the mid-1950s—but also manipulating them. Moreover, unlike Drexler's approach, which involved the complicated process of bioengineering new protein tools—something experts like Kevin Ulmer came to recognize would be far more expensive and time-consuming than he and eager journalists had anticipated—STMs already existed. Even more, scores of researchers at universities and corporate labs were rapidly improving their performance and versatility. Schneiker saw the STM as a general-purpose tool in a manner similar to how Drexler imagined his assemblers.

In 1985, Schneiker wrote a series of papers—short, prescient, but ultimately unpublished. In them, he speculated that the STM could be used not just to "see" atoms but also to "move" them. In February 1985, for example, he predicted that "with suitable modifications, the STM can be used to directly manipulate individual atoms and molecules."[101] Schneiker, who was auditing classes at Caltech at the time, decided to honor the Caltech physicist he so admired by calling the hypothetical machine tools based on the STM "Feynman machines." (Richard Feynman, for his part, appears to have been unaware of the new microscopes until Schneiker told him about them.)

Throughout 1985, Schneiker tried to drum up interest in "Feynman machines" and nanoscale engineering in general. He wrote letters to a variety of potential patrons, including the U.S. Army, Texas Instruments, and Silicon Valley venture capitalists.[102] The polite refusals he received came with the expected reasons: Schneiker, like Drexler, presented plausible-sounding ideas. However, besides not having a PhD or an affiliation with a university or major firm, he did not provide the details or experimental evidence that could point the way toward a specific program of research.

Nonetheless, Schneiker's extrapolations of the STM's capabilities anticipated the first deliberate modifications of matter at the atomic scale using an STM by at least a few years. In October

1985, after corresponding with Schneiker, Paul Hansma, a physics professor at the University of California in Santa Barbara who helped improve the original STM design, wrote Feynman. Hansma told the Caltech physicist that he was "excited about the possibility, as suggested by Conrad Schneiker, of using a tunneling microscope as a miniature robot arm."[103] A year later, when Binnig and Rohrer jointly received the Nobel Prize in Physics for their design of the STM, they noted that the instrument might someday be used, referencing an unpublished paper by Schneiker, as a "Feynman machine" to "ultimately handle atoms."[104] That day came quickly. In January 1987, researchers at Bell Labs altered a semiconductor's surface at the atomic scale with an STM. These results suggested a properly equipped STM might one day be able to "modify materials on the smallest possible scale" leading to a wide range of uses including "high-density information storage" and even the "purposeful transformation of genetic material." Such applications would require substantial improvements to the STM. Nonetheless, the Bell Labs team expressed their pleasure at having shown "that man can now manipulate a few chosen atoms for his own purposes."[105]

Lab results such as these confirmed Schneiker's insight that the STM could be a tool for moving as well as seeing atoms and molecules. He returned to Tucson to explore the possibilities of "Feynman machines" further, taking a joint research assistant position at the University of Arizona's Optical Sciences Center and its medical school. His main collaborator was Stuart Hameroff, a physician and anesthesiologist studying mechanisms of consciousness. In the mid-1980s, Hameroff was especially interested in the role played by microtubules—cylindrical assemblages of proteins found in neurons—in intelligence and thought. He wondered, for instance, if these biological features allowed the brain to process and store information like a computer. The STM, Hameroff thought, could allow scientists to more closely study microtubules and other biological structures. Schneiker helped build a series of STMs and, over the next few years, he and Hameroff wrote about the instrument's potential for future computing and biomedical applications. They even described what they called a "NanoTechnology

Workstation" that combined an STM with traditional optical microscopy to image and manipulate molecules.[106]

As the name of their proposed instrument ensemble suggests, by the mid-1980s both Drexler and Schneiker had started to use the word *nanotechnology* as a general term to describe their ideas. However, neither of them could claim credit for the neologism. Since at least the 1950s, researchers had used "nano" as a prefix to represent one billionth of something, as in a nanometer or nanowatt. In August 1974, a relatively obscure sixty-two-year old Japanese mechanical engineer named Norio Taniguchi surveyed the array of techniques engineers could use for high-precision manufacturing.[107] Speaking at a conference in Tokyo, Taniguchi labeled this ultraexact form of machining "Nano-Technology." Like Feynman's "Room at the Bottom" talk, Taniguchi's term disappeared into library stacks until nanotech advocates rediscovered it years later. In any case, Taniguchi's concept represented something entirely different from what either Drexler or Schneiker had in mind.

Etymology aside, by 1986 Schneiker had prepared his own book manuscript. Titled "NanoTechnology with Feynman Machines," his manuscript presented a vision of how nanotechnology might be achieved, one that differed significantly in strategy and style from what Drexler was working on.

Drexler's path to the future would be built with biotechnologists' test tubes, petri dishes, and electrophoresis gels. He imagined molecular manufacturing as a revolutionary new technology based on the functions of ribosomes, enzymes, and living cells. Engineered protein machines were theoretically possible, he argued, but their realization would take at least a few decades. Nevertheless, as Drexler often argued, the proof of concept for his vision already existed—life itself.

In contrast, Schneicker's version relied on cutting-edge microscopes. He envisioned a path to the nanotechnological future that was much more incremental. He was struck not just by Feynman's vision but also by the decades of patient technological development that had been required to achieve the first rudimentary nano tools like the STM. Moreover, Schneiker based his ideas much more on the microelectronics industry's steadily improving ability as it followed Moore's Law and manufactured ever-smaller features on computer chips.

How they presented their ideas differed as well. Schneiker's version appeared more as a catalog of the previous research he considered to be the essential foundation for future nanotechnologies—he even got Feynman's permission to include a reprint of "Plenty of Room at the Bottom" in his book. Schneiker's single-spaced bibliography ran to more than fifty pages, and he filled his book with long passages excerpted from the writings of renowned scientists like Freeman Dyson and John von Neumann.

In contrast, Drexler's taut writing possessed a more expansive, even extravagant quality that linked technological speculations to politics, societal concerns, and broad depictions of the future. In his book manuscript, Drexler gave far more attention to how the future might be affected by nanotechnology, for good or ill, than to specific research accomplished or underway that would help ground his vision for molecular technology. Of the two, Drexler's was the more complete, well-crafted, and eloquent vision. Yet it required tools that did not yet exist and seemed years, if not decades, from realization.

Given the different paths they advocated for achieving routine atomically precise manufacturing, it's not surprising that a rift developed between Schneiker and Drexler. Schneiker came to see Drexler as wanting to promote himself as much his technology and found the dearth of details in Drexler's writings increasingly bothersome. "He was never forthcoming with answers," Schneiker later recalled.[108] Similar criticisms could be made of Schneiker's work as well, as both men engaged freely in speculation about future technologies and their implications.

Drexler, for his part, stuck to his belief that "protein machines" would lead to the nanotechnology he envisioned. He perhaps also considered Schneiker as a rival for ownership of an idea he believed would be transformative. There also was the sense of a race—both men, after all, were working on book manuscripts about nanotechnology. But, here, the advantage went to Drexler. He eventually secured a contract with Doubleday, a New York–based publisher, and enjoyed the support of well-connected advisers like Marvin Minsky, whereas Schneiker would do his own publication and distribution.

To be sure, Drexler didn't seem to appreciate Schneiker's insistence that STMs offered the best path to the nano future. Drexler took the position that molecular technology would be revolutionary but, misused or in the wrong hands, potentially hazardous. As he told a group in 1985, "I looked at this [STMs] in '82, and I said, 'Oh. No! This just might be a shortcut to a technology that strikes me as being very dangerous,' so I said nothing about it. . . . STM might be a shortcut to the same goal, but if so, it's not obvious how it will work."[109] For these reasons, and perhaps others, Drexler gave the STM short shrift in the early years of formulating his ideas.

To Schneiker, such concerns appeared contrived. He even claimed Drexler, Christine Peterson, and Mark Miller had tried, while attending a space development conference in Washington, DC, to dissuade him from even talking about the potential of STMs. Perhaps this was out of fear that some sort of super-STM might lead to dangerous technologies. Or maybe it reflected Drexler's eagerness to avoid another round of revisions as his book finally moved into production. To resolve the dispute, a provisional "science court" was convened, which Arthur Kantrowitz presided over. After hearing both sides, Schneiker recalled, Kantrowitz "vigorously argued for public discussion of scanning tunneling microscopes and related technologies."[110]

To some degree, the quarrels between Drexler and Schneiker resembled a classic priority dispute in which scientists argue about credit, recognition, and accuracy. Of course, neither Schneiker nor Drexler had made a specific laboratory discovery. Likewise, neither of them could claim credit for coining the term *nanotechnology*. And, of course, speculative ideas about tiny machines and the power to precisely control and manipulate matter at the molecular scale had been around at least since von Hippel and Feynman spoke and wrote on the subject in the 1950s. Instead, their disagreements were partly a matter of principle, personality, and ambitions. But, like the friendly disagreements Gerard O'Neill had with Freeman Dyson and other space settlement advocates, they were also about the future—namely, what was the "right" path one should take in order to see nanotechnology realized?

To be sure, the dispute between Schneiker and Drexler was about next to nothing. As an identifiable field of engineering, nanotechnology was practically nonexistent in 1985. As a word, *nanotechnology*, coined years before by an obscure Japanese engineer and forgotten (neither man was aware of Taniguchi's paper at the time), had almost no salience to the public or researchers. In the mid-1980s, both Drexler and Schneiker were laboring at the margins of obscurity. Most of the sniping between the two men appeared, of all places, in *Cryonics*, a small-circulation monthly publication put out by the Alcor Life Extension Foundation. As a result, their discussions about what was and wasn't possible with nanotechnologies went largely unnoticed. Yet this dispute makes plain that, even at its inception, nanotechnology was contested terrain, right down to what it was (and wasn't) and how it might be achieved. As we'll see in the next two chapters, in a very real sense, the *longue durée* of disagreements over nanotechnology's evolving definitions and demarcations—disputes that began in the last years of the Cold War and continued into the post-9/11 era—*is* its history.

Conrad Schneiker never lost his interest in nanotechnology and continued to support himself via various engineering-related activities. Over the next three decades, he shuttled back and forth between graduate study and corporate software development. Throughout his peregrinations, he remained focused on the necessary "intermediate steps" to realize his STM-based approach to nanotechnology.[111] He eventually received several patents for modifications he proposed to improve the STM's basic design and imaging capabilities.[112]

To be sure, Schneiker was not a visioneer cast from the same mold as Drexler or Gerard O'Neill. He had a technical background as well as a willingness to build tools necessary to advance his vision for nanotechnology. But he did not strive to popularize his ideas to a wide audience, build a large community of supporters, or engage the interest of policy makers. He never achieved widespread recognition for envisioning, correctly at times, what STMs might be able to do. But launching a technological revolution was not his primary goal.[113]

Drexler, on the other hand, positioned himself as a source of engineering ideas for his particular view of nanotechnology as well as someone who would help promote and popularize it. Regardless of the initial debates over whether nanotechnology would develop via test tubes or electron microscopes or whether the public would see nanotechnology emerge via a sudden revolution or more measured change, the fact was that, after 1986, nanotechnology lost its invisibility. Drexler's *Engines of Creation* was published, and the young visioneer became the public face of the nanotechnological future.

CHAPTER 6

California Dreaming

||||||||||||||||||||||||||||||||||||||||||||||||||||||||||||||||||||||||||||||||||||||||||||||||||||||||||||||||||||||||||||

As I've said many times, the future is here. It's just not very evenly
distributed.
—William Gibson, science fiction author, interview on
National Public Radio, November 30, 1999

Only a few years after nanotechnology had blossomed into a
global research initiative that consumed billions of government
and corporate dollars, one well-placed observer claimed the field,
if indeed nanotechnology could be delimited to such a thing, was
experiencing an "existential crisis."[1] But how could the "brave and
wondrous new world" of nanotechnology already be in such a
perilous state?[2]

One answer was that nanotechnology *wasn't* new. Quite the opposite. Despite the oft-used revolutionary rhetoric, its foundations
were rooted in Cold War–era microelectronics and molecular biology. Scientists like von Hippel and Feynman proposed a future in
which precise manipulation of the molecular realm was attainable.
Early nano visionaries like Conrad Schneiker and Eric Drexler
built on these imaginative ideas, adding concepts drawn liberally
from two decades of advances in biology, microelectronics, and
computer engineering. Schneiker had even deliberately used the
term "Feynman machine" to "emphasize the fact that this field
wasn't just something that came up, that these ideas weren't brand
new."[3] Drexler, meanwhile, proceeded to popularize what he only
gradually, and somewhat reluctantly, called nanotechnology.

The existential quandary also emerged because there was no
single entity to which one could point and say "*That* is nanotechnology." Nanotechnology wasn't like Thomas Newcomen's steam
engine or Henry Ford's assembly line. It was instead a broad and

enabling set of technologies for production and a rubric of ideas that purported to contain the seeds for radical social and economic change. Seen this way, it more closely resembled the scientific management of Taylorism, the early-twentieth-century principles of "scientific management" that proposed how assembly lines and factories should be organized. Much closer to an ideology than to actual nuts-and-bolts engineering, Taylorism (and nanotechnology) helped unite a community of adherents and boosters in pursuit of a new future for industry and society.

Like Ford's assembly lines, nanotechnology appeared during and drew on an extraordinary period of American technological and industrial change. Following the end of World War Two, a powerful new social contract emerged between universities, business, and government. The participants in this "triple helix" framed scientific research and technological development as essential components of America's global power during the Cold War.[4] Established disciplines like physics and chemistry grew handsomely while new fields like materials science, molecular biology, and computer science emerged and thrived. Although nanotechnology today has the cachet of cutting-edge science and engineering, its proponents foraged extensively on the many intellectual and technological frontiers that researchers had opened during this prolific period.

Even among its early supporters, nanotechnology possessed a certain plasticity. This flexibility allowed proponents such as Schneiker and Drexler to interpret and imagine it in diverse ways. Following Drexler's successful popularizing, "nanotechnologists" from a wide array of academic disciplines—biology, physics, chemistry, materials science, engineering, each with its own argot, professional societies, conferences, and journals—embraced a broad ensemble of tools and techniques. Meanwhile, the visioneering of people like Drexler in the late 1980s and throughout the 1990s helped create excitement and high expectations among businesspeople and journalists.

The confusion over what nanotechnology was and who was a nanotechnologist (a label that appeared with any regularity only after 1990) was compounded when funding for and public recognition of nanoscale research started to dramatically increase. Established researchers were tempted to rebrand existing work as "nano," while graduate students and younger scientists embraced the novelty of a seemingly new field and the resources it offered. Nanotechnology became a hybrid, an interdisciplinary arena where the search for scientific understanding coexisted uneasily with government and corporate desire for new products and economic competitiveness.

Fundamentally, nanotechnology's own history was the catalyst for its existential angst. Vastly different interpretations of nanotechnology, both as a research program and as a vision for the future, emerged between Drexler's early publications and the launching of a major national initiative in the United States two decades later. To complicate things further, just as enthusiasts co-opted Gerard O'Neill's ideas, Drexler's visioneering took on a life of its own. Consequently, one must distinguish between Drexler's own ideas (which themselves evolved); variations of his ideas promulgated by "Drexlerians" (a community Drexler professed some misgivings toward even as he helped foster it); and the ways journalists, artists, and science writers presented nanotechnology to a wider audience.[5] As a result, opportunities for misunderstanding were rife.

As had been the case for microelectronics and biotechnology, the first wave of articles in popular magazines assumed an optimistic tone. Even *Reader's Digest*, a bastion of populist ideas, spoke of the coming age of "Tiny Tech."[6] Aided by favorable media coverage, Drexler and Christine Peterson assembled a diverse community devoted to radical concepts for molecular machines. Life-extension advocates, software programmers, computer engineers, business executives, and the international techno-cognoscenti came out as early supporters of nanotechnology. Although this publicity generated interest and support for Drexlerian nanotechnology, it also set the stage for a stinging backlash from some leading figures in the mainstream scientific community.

Popularizing Nanotechnology

As Gerard O'Neill had also discovered, getting published wasn't easy. Drexler recalled that his book draft—after it was "ripped to shreds" by friends and colleagues, rewritten, and critiqued anew—met rejection from at least ten publishers. He then asked MIT's Marvin Minsky to provide a foreword. Finally, "possibly by coincidence, but possibly not," an agent for Doubleday agreed to publish what Drexler titled *Engines of Creation*.[7] Anticipating a print run of "only 7,000 copies or so" (excellent figures for an academic book but modest for a trade book), Drexler reached out to people he knew from the pro-space and life-extension communities to drum up interest so his publisher would not judge *Engines* as "just another non-fiction book by a non-famous author."[8]

Appearing in mid-1986, *Engines* eventually sold over 100,000 copies. Starting with a Japanese edition in 1992, it was also translated into several other languages.[9] Gerard O'Neill even provided his former research assistant with a dust jacket blurb. Drexler's clear and accessible writing style aided the book's popularity. The stale references to history and political philosophy were gone, as were the overt libertarian-inflected references that bogged down earlier drafts. Given the imagined power of molecular assemblers, Drexler even acknowledged that some government oversight and funding would be necessary for their beneficent implementation.[10]

Ironically, given how Drexler would become indelibly associated with nanotechnology, the word itself didn't appear in the book's original title. Doubleday initially published it as *Engines of Creation: Challenges and Choices of the Last Technological Revolution*. Only with later editions did the publisher change the subtitle to *The Coming Era of Nanotechnology*. The switch is revealing. The initial title reflected Drexler's original intent to offer, as O'Neill had with his *High Frontier*, a broad vision of the technological future. The new version reflected the beachhead that "nanotechnology" was beginning to secure among technology enthusiasts, business leaders, and the research community.

Reactions to *Engines* were generally positive. In the *New York Times*, Terence Monmaney, who wrote about science for *Smithsonian*, called it a "clearly written, hopeful forecast" that expressed an "unembarrassed faith in progress through technology." Words like "faith" would later assume a different meaning when scientists and journalists critiqued the messianic aura that enthusiasts ascribed to Drexler and nanotechnology. Countering the pessimism of *Limits to Growth*, *Engines* described a cornucopian future decades hence of "no pollution or disease" and "limitless energy and food."[11] "Future-reading at its best," tool-guru J. Baldwin concluded in the *Whole Earth Review*.[12]

Although Drexler began his book with biology-based descriptions of how biological engineering could lead to general-purpose assemblers and nanotechnology, *Engines* plunged ahead to other topics that technology enthusiasts—imagine a typical *Omni* reader—would find appealing and provocative. Robotics, artificial intelligence, cutting-edge computing, space exploration, novel information technologies, life-extension research, and technological forecasting all appeared in *Engines*'s pages. In fact, we might imagine Drexler himself as an "assembler" as he engineered a social network that connected ideas for molecular technology to many other radical technologies and proposed their synergistic convergence.

Even before *Engines* appeared, a community of enthusiasts for nanotechnology had started to coalesce around Drexler and his ideas. As was the case for O'Neill's space colonies, Drexler raised initial interest among college students, especially on MIT's campus. In January 1985, he used the school's Independent Activities Period, held between semesters, to organize a series of nanotechnology-related lectures. In addition to his own talk, called "Limits to the Possible," sympathetic MIT faculty such as Marvin Minsky spoke.

With Drexler's encouragement, MIT students formed a Nanotechnology Study Group. Over the next several years, its members held retreats and invited speakers for annual symposia. Not all the talks were about science and engineering. In 1987, David Friedman—libertarian theorist, medieval reenactor, and son of free-market advocate Milton Friedman—spoke. Group members, many associated with MIT's Artificial Intelligence Laboratory, assembled an ever-growing set of articles into a "Nanotechnology Notebook" that MIT's engineering library kept on reserve. They also made a press kit and set up an Internet-based discussion forum—a relative novelty in the mid-1980s—that people affiliated with MIT could access.[13]

As was the case with O'Neill's supporters, this community organized itself primarily around ideas and texts such as Drexler's book as well as others addressing molecular computation, cognitive science, AI, and politics. These included, for example, Minsky's *Society of Mind*, a classic work on human cognition, and Hans Moravec's *Mind Children*, which extrapolated Moore's Law to imagine a world in which robots, computers, and people interact and merge. The nascent nanotech community not only read and discussed others writings about technology but also produced some of their own. For example, one MIT engineering student wrote a series of essays on nanotechnology that won a national prize in Honeywell's Futurist Awards Competition.[14] Over the next few years, nanotechnology study groups replicated in places like Tucson, Palo Alto, and Seattle. Not all groups functioned as harmoniously as MIT's—one member of Berkeley's "NanoPIRG" group critiqued its "circus-like atmosphere" where "primadonna nerds and fringe intellectuals" dominated discussions.[15]

If nanotechnology was going to become as important as Drexler believed, however, then it was necessary to expand beyond campus discussion groups. In the spring of 1985, Drexler and Christine Peterson left MIT's familiar environs for the San Francisco Bay Area. Their destination made sense for a number of reasons. Like Boston, the Bay Area was home to top-notch schools. Many of their friends were already living there or getting ready to relocate. This included James Bennett, a former L5 member who, with George Koopman, had recently started a privately owned venture called American Rocket Company. Job advertisements for computer experts were plentiful, and Mark Miller moved to California at about the same time as Drexler. He recalled, "We were running up tremendous long-distance bills talking to each other. We needed to rendezvous and form our own community."[16]

Given the Bay Area's rich ensemble of microelectronics, computer, and biotechnology firms—all key elements in nanotechnology—coupled with the region's overall entrepreneurial ethos, the West Coast was a magnet for someone with Drexler's interests. Moreover, it was a region built on an ideal of continual and sometimes revolutionary technological change. Soon after they arrived, Drexler secured a temporary visiting researcher position in Stanford's Computer Science Department—no pay but also no duties, and an elite affiliation—and he and Christine Peterson considered their next step.

Silicon Valley's technological ecosystem thrives on entrepreneurs and small start-ups that spin off from larger firms or the region's universities. Only a few such ventures make it big. Drexler and Peterson set out to create a different sort of species—a nonprofit and nongovernmental organization that would "help society prepare for new and future technologies" by promoting sound policies and an informed public.[17] Nanotechnology would be a central topic, but they also had plans to consider developments in areas such as artificial intelligence, life extension, and space development. They named their organization the Foresight Institute. By the end of 1987, the IRS had approved Foresight's tax-exempt status with Drexler as president, Peterson as secretary and treasurer, and space entrepreneur Bennett rounding out its initial board.

All three of them were eager to avoid the mistakes they watched the pro-space movement make. "What I was picturing was a kind of a replay of the Gerry O'Neill space movement scenario," Drexler recalled, "with me sitting in the O'Neill position, and not wanting an L5 Society to happen, which is to say some freewheeling agglomeration of enthusiasts on stage." Instead, he saw Foresight occupying an "ecological niche that would be more benign . . . more muted and controlled."[18] Bennett remembered that L5 and the pro-space movement had "engendered lots of mass participation," which also drew in some "eccentric individuals." "We didn't want to be as mass-movement oriented as L5," he recalled, "because there was less control of the message."[19]

However, the message, as O'Neill had learned, could be slippery to control even under the best circumstances. At first, the Foresight Institute could only offer Drexler's own writings on nanotechnology. By 1990, however, a whole array of "Drexlerian" interpretations from other enthusiasts and journalists had appeared, replicated, and spread. Drexler tried to quell what he called "bogosities" about nanotechnology, especially those that violated physical laws, but the swell of interest sometimes startled him. One scientist who attended an early conference organized around Drexler's ideas recalled that "Eric seemed a little freaked with the 'true believers' in the audience" and observed that "the cultish people who gathered around him did not do science any favors."[20]

Starting with about a hundred potential prospects, many carefully drawn from L5's ranks, Foresight grew to a few thousand dues-paying members.[21] Although Foresight was generally positive about nanotechnology, Drexler tried to position himself not as an unabashed cheerleader but as someone offering objective information about a potentially powerful technology that could be used or misused. Drexler and Peterson were also keen to avoid the political fragmentation that bedeviled the space movement. "The worst scenario I imagine is this discussion becoming polarized," he told one journalist in 1988, "where you have a Lyndon Larouche on one hand screaming for weapons development and a Jeremy Rifkin on the other, blathering about a moratorium on all advances. We're trying to plant a flag somewhere in the middle."[22]

Two high-profile articles, one in *Omni* and another in *Scientific American*, helped Drexler plant this flag and increase public awareness about nanotechnology. In comparison to Drexler's own essay in *Smithsonian*, these treatments, by including more technical details, appealed to science and technology buffs. Besides bringing wide attention to "nanotechnology," which was still a relatively new term, these pieces offered descriptions, images, and predictions that, for better or worse, defined the topic for years to come.

Omni, by now a mature magazine read by more than million people every month, made nanotechnology its main story for its November 1986 issue. *Omni*'s cover showed a person silhouetted beneath a glowing cube filled with clouds, its surfaces interlaced with blue neon streaks while above there was the bold headline "Nanotechnology: Molecular Machines that Mimic Life." Inside, a feature piece titled "TinyTech" focused on Drexler as well as the activities of MIT's Nanotechnology Study Group.[23]

Besides presenting the basic view of future nanotechnology achieved via biotechnology and protein engineering (with Drexler likened to Johnny Appleseed, Cassandra, and Paul Revere), the *Omni* article included a concept that figured prominently in popular discussions of nanotech for years to come. This was the idea of "gray goo," a term Drexler had first used in *Engines* to describe

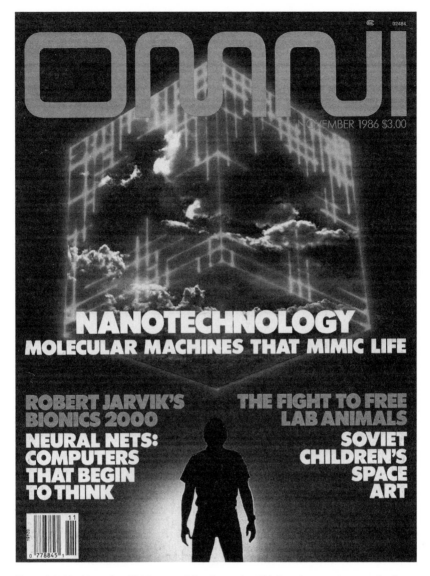

Figure 6.1 The November 1986 cover of *Omni* magazine. This issue's cover story was the first major journalistic account of Drexler and his conceptualization of nanotechnology. (© Dale O'Dell 1986.)

malfunctioning nanomachines. Replicating unchecked under the right circumstances, "tough, omnivorous nanobacteria" could, according to *Omni*, potentially "outcompete real bacteria" and create a serious environmental hazard. One extreme outcome could be "reducing the biosphere to dust in a matter of days."[24]

Such a hypothetical scenario resembled what opponents of recombinant DNA had warned about in the 1970s. Drexler, of course, had been a student at MIT when communities like Cambridge enacted regulations on rDNA research. Although he gave the "gray goo" scenario only a few brief mentions in *Engines*, it was one reason Drexler believed nanotechnology's development needed to be carefully managed. Gray goo later assumed an almost eschatological role in science writers' depictions of nanotech. Consequently, images of out-of-control "nanobots"—a term that Drexler himself derided but one that entered the general lexicon nonetheless—became entrenched in public imaginings of nanotechnology.

Scientific American, which had circulation figures comparable to *Omni*'s, promulgated the idea of the "nanobot" further. In January 1988, A. K. Dewdney, a Canadian scientist who wrote about computers and mathematics, introduced nanotechnology to the magazine's scientifically literate audience.[25] He did this by highlighting a design for a hypothetical "nanocomputer" that Drexler had proposed. Dewdney's essay also showed a "nanomachine swimming through a capillary" as the sort of device that might be controlled with molecular computers. Science fiction fans, of course, recognized this image as a descendant from the 1966 film *Fantastic Voyage*, in which Raquel Welch and her fellow explorers in a miniaturized submarine are injected into the bloodstream of an injured scientist. Feynman himself had even speculated about tiny robotic surgeons that could be swallowed. Fantastic to be sure, *Scientific American*'s treatment helped create an expectation among some hopeful readers that nanotechnology might soon lead to cell-repair machines.[26] Although Dewdney remained unconvinced by Drexler's ideas, calling him "an informed speculator" and a "technologist without a technology," the article generated hundreds of inquiries about the Foresight Institute.[27]

Looking beyond widely read magazines like *Omni* (which featured Drexler again in 1989) and *Scientific American*, one finds a huge corpus of science fiction that incorporated nanotechnology into its plots. Some of these works were read only by small groups of devoted fans. Other nano-based stories found audiences of millions. In 1989, for example, the television series *Star Trek: The Next Generation* introduced Trekkies to nanotechnology. An episode called "Evolution" featured the accidental release of sentient "nanites" (i.e., nanobots) into the computer core of the starship *Enterprise*. Replicating unchecked, they sabotage the ship until a truce between them and the crew is reached. The *Star Trek* episode was just one example of a larger pattern as "assemblers" and "nanobots" became oft-used, even overused, plot devices for fiction and film.[28] By 1999, more than three hundred books, stories, television shows, and films had incorporated some radical type of nanotechnology.[29]

Like Chesley Bonestell's stunning paintings of spacescapes and the space-themed films from the 1950s, these works helped influence the public's imagination of nanotechnology and what its realization could achieve.[30] This "nanovision," which dissolved the "membrane between the technological present and the nanotechnological future," was not limited to science fiction authors or Drexler and members of Foresight.[31] Scientists themselves eventually began to use science fiction and futuristic imagery to discuss and draw attention to their nano research.[32] Meanwhile, magazines as culturally removed from one another as *Interview*, the celebrity-laden "crystal ball of pop" started by Andy Warhol, and *Parade*, a national Sunday newspaper supplement, presented interpretations of Drexlerian nanotechnology. By depicting diverse and radical futures brought about nanotechnology, this science fiction and media coverage created its own version of nanotechnology. To be sure, it certainly influenced political and journalistic discussions about it throughout the 1990s and beyond.

Nanotech's Early Supporters

When I visited the Foresight Institute in 2007, I was shown file folders stuffed with news clippings and magazine articles, most with dates spanning a roughly ten-year period starting in the mid-to late 1980s. The stacks of newsprint attested to the intense interest in Drexler and his visioneering of nanotechnology that blossomed during this time. Peterson and Drexler helped amplify media interest with the information-packed newsletters they sent out regularly to Foresight members and other technology enthusiasts. By providing updates on Drexler's activities, the Foresight newsletters also helped promote nanotechnology. One issue, for example, listed some twenty institutions at which he had spoken. Another described a new undergraduate course (Nanotechnology and Exploratory Engineering) he was able to offer through Stanford's Computer Science Department to students interested in the "fundamentals of nanotechnology and the engineering of molecular devices."[33] The net effect was to convey both the importance of nanotechnology and Drexler's direct connections to it.

By suggesting recently published books about technologies that might interest members, the newsletters and other breviaries that Foresight distributed helped anchor a growing community around other canonical texts besides *Engines*. An electronic bulletin board service—sci.nanotech—started by a graduate student in computer science at Rutgers University in 1988 offered another avenue for the community. If they weren't building molecular assemblers, nano aficionados could at least debate their underlying science and potential implications, intellectual exercises similar to what space colony advocates had once engaged in.

Drexler occasionally wrote short pieces for Foresight about his own work. His interests often delved into the realm of computer science. He evaluated, for example, the scientific potential of molecular modeling software available for the Macintosh personal computer. In another issue, he discussed the premise of "hacking molecules"—not in the sense of "computerish juvenile delinquents" but how "inventive technologists" could make computer modeling "jump through hoops" and help foster "amateur nanomachine design."[34]

Besides helping spread Drexler's ideas, Foresight's newsletters also offered a clearinghouse for news about nanotech-related discoveries made by researchers at university or corporate labs in areas such as electronics, micromanipulation, and protein engineering. Highlighting scientific advances relevant to nanotechnology helped create the impression that progress toward molecular manufacturing was happening apace. It also positioned Foresight as a key source of information about nanotechnology while indirectly linking it to places where actual benchtop research was happening.

Despite its imposing name, Drexler and Peterson started operating the Foresight Institute on a lean budget out of their modest Spanish-style stucco house near Stanford's campus. Members' contributions provided some income for Drexler and Peterson, and they worked assiduously with Bennett to court donors with deeper pockets. Silicon Valley offered good opportunities to make connections with wealthy venture capitalists who had done well in computer or biotechnology companies.[35] During Foresight's early years, the organization also received considerable support from telecommunications entrepreneur Barry Silverstein. Silverstein learned of Drexler's work via *Omni* magazine and quickly bought a copy of *Engines*, which he read and shared with colleagues. According to one writer, Silverstein's support for Foresight eventually reached into the six figures.[36]

Membership in Foresight proved especially attractive to people who worked in the major industries that called coastal California home—aerospace, biotechnology, microelectronics, software and programming. Other people came from less visible communities in the area's technological ecosystem. One early group especially attracted to nanotech was life-extension advocates.

Potential approaches for extending one's life span, of course, had long been a regular theme in science fiction as well as future-oriented magazines like *Omni*. For some people, this perhaps meant longevity via caloric restriction. Others looked to a more radical solution such as cryonics—the low-temperature preservation of a person's body or brain in the hopes that future medical advances might be able to bring about revival. Some people even talked about the more extreme possibility of uploading one's memories into a computer and eschewing a corporeal existence altogether. But, whatever the path, California hosted one of the world's largest communities devoted to life extension, and many of these people had ties to the Golden State's various high-tech companies.

The idea of people going into cold-induced suspended animation has long been a staple of fiction and films, especially those depicting long-term travel in space. Even rocket pioneer Robert Goddard speculated about so-called "generation ships" that might one day carry people, their life functions suspended, far out into space. "It has long been known," he wrote, "that protoplasm can remain inanimate for great periods of time, and can also withstand great cold, if in the granular state."[37] In the 1970s, L5's newsletter often carried advertisements about life-extension-themed books and meetings that hinted at the overlap of interests among the group's members.

In the 1960s, ideas about life extension were discussed by some reputable scientists at established institutions, often in connection with space exploration or technological utopianism in general. For example, in 1960 two medical researchers presented the possibility of lowering a person's metabolic rate for the Space Age's expected voyages.[38] Six years later, *Physics Today* published Gerald Feinberg's "Physics and Life Prolongation." A scientist at Columbia University whose theoretical research included hypothetical faster-than-light particles called tachyons, Feinberg wrote in his article that "freezing and storing at low temperatures might lead to many new potentialities for the human race."[39] The 1960s also saw the appearance of Robert Ettinger's *Prospect of Immortality*, published by Doubleday (also the publisher of Drexler's *Engines*) and included as a Book-of-the-Month Club selection. While publications such as these helped foster a wider audience, most mainstream scientists still gave cryonics a chilly reception.[40]

Nonetheless, cryonics became front-page news in January 1967 when reporters announced that a Californian was the first person frozen for a "future revival experiment." Contrary to urban legends, it was not entertainer Walt Disney but rather James H. Bedford, a seventy-three-year-old former psychology professor from Glendale.[41] After doctors pronounced Bedford dead (his last words reportedly were "I'm feeling better"), members of the Cryonics Society of California injected him with an anticoagulant and placed him in a metal "cryo-capsule" filled with liquid nitrogen. This sort of experiment had already been anticipated by science fiction writers such as futurist Frederick Pohl, who told insurance industry executives to brace themselves for the "$30 trillion market of the future" by writing policies to cover cryonic suspension.[42]

Soon after its first "experiment," the Cryonics Society of California announced plans to open a twenty-person storage facility in the Los Angeles area. A much larger facility was anticipated in Barstow for patients willing to pay up to $20,000 for the procedure.[43] New organizations popped up to support the growing community. For example, Fred Chamberlain, an employee at Pasadena's Jet Propulsion Laboratory, and his wife Linda started a company called Manrise and self-published one of the first manuals for cryopreservation.[44] The Chamberlains went on to form the Alcor Society for Solid State Hypothermia, naming their group after a star in the Big Dipper constellation that had long been a test of a person's vision and focus.[45]

But, at first, exactly how "cryonauts" might experience a high-tech revival was largely left unexplained. A potential answer to this key question emerged in the mid-1980s: nanotechnology.

Drexler's first articles about nanotechnology (as well as an entire chapter in *Engines of Creation*) discussed how future molecular devices might repair frozen or damaged tissue. As a result, cryonicists were one of the first communities to enthusiastically embrace what were among the most speculative ideas about nanotechnology. Both Drexler and Schneiker advocated cryonics and joined groups such as Alcor. Meanwhile, L5 cofounder Keith Henson became a "convert to nanotechnology" after reading drafts of *Engines* and promoted it in Alcor's monthly newsletter *Cryonics*.[46] In fact, the small community of cryonicists discussed "molecular engineering" as "the kind of technology which will likely be required to revive us" well before *Engines of Creation* even appeared, making them among the first people to consider its revolutionary potential.[47] For years, the pages of *Cryonics* were rife with discussions of nanotechnology while Drexler and others who made bold predictions for nanotechnology spoke routinely at cryonics conferences.[48]

More than a few scientists from the mainstream of academic research supported the idea of life extension via nanotechnology. For example, Gerald Feinberg and Marvin Minsky advocated cryonics and also were listed as members of the Foresight Institute's advisory board. Drexler's ideas, in other words, helped form a bridge of sorts between molecular engineering and life extension, just as he had also helped connect futuristic space exploration ideas with nanotechnology. The enthusiasm life-extension advocates had for the more radical versions of nanotechnology, however, did not help Drexler win converts among some mainstream academic scientists and became one more reason for his marginalization in the 1990s.

Californians made up the largest fraction of Alcor's overall membership which tripled throughout the late 1980s in step with growing publicity about nanotechnology.[49] Like space entrepreneurship and molecular engineering, the possibility of life extension via cryonics found an especially warm reception among the Silicon Valley's high-tech culture. The *San Jose Mercury News* profiled several of these "nerds on ice" for whom "technology is everything." "Techies," one of them explained, would "accept and dive into things" before they became widely accepted.[50] Drexler and several of his friends signed up for cryonics, and when his friend Phil Salin passed away from cancer in late 1991, he was cryopreserved via cephalic isolation (i.e., his head was removed and frozen) at an Alcor facility.[51]

The interest in cryonics among computer engineers guides us to another community that found Drexler's ideas for nanotechnology especially appealing: the Bay Area's high-tech communities, especially Silicon Valley's extensive software and computer businesses. This was a community Drexler had direct links to via many of his friends as well as a long-running computer project called Xanadu.

Xanadu had originated with computer guru Ted Nelson. In 1965, nine years before he wrote *Computer Lib/Dream Machines*, Nelson coined the term "hypertext." For him, this meant a "body of written or pictorial material interconnected in such a complex way that it could not conveniently be presented or represented on paper."[52] A few years later, Nelson developed what became an evolving design for just such an all-encompassing digital data system and christened it Xanadu.

Of course, the idea of a universal library was an age-old dream. For example, in a famous 1945 essay called "As We May Think," Vannevar Bush famously described a hypothetical device, akin to a microfilm reader, called a "memex," whose screens and cameras would allow users to rapidly access and annotate documents.[53] Taking some inspiration from Bush—Nelson once gave a talk called "As We Will Think"—he imagined a proprietary computer system with which one could see and make links between a wide range of media including articles, film, and pictures.[54] As he imagined in *Computer Lib/Dream Machines*, Project Xanadu offered "new freedoms through computer screens."[55] Nelson even composed an advertising jingle to promote his hypothetical information-distribution-for-profit franchises, which he called "Silver Stands":

> The greatest things you've ever seen
> Dance your wishes on the screen
> All the things that man has known
> Comin' on the telephone—
> Poems, books and pictures too
> Comin' on the Xanadu.[56]

Several of Drexler's friends, including Mark Miller, found Nelson's ambitions irresistible and joined the Xanadu project. Drexler himself did some consulting work for it. For years, the Xanadu team devoted themselves with an almost religious intensity to engineering the digital tools they needed and raising money to keep the project afloat. However, even as the World Wide Web was emerging, Xanadu collapsed under the weight of its own ambition and complexity.

Although Xanadu failed to materialize as reality, its underlying principles had appealed to Drexler for years. Computer-driven tools like hypertext could help people, for example, "gather and organize knowledge" and also ease its evolution and its evaluation. Drexler predicted that hypertext would eventually allow critics of misinformation to "plant their barbs firmly in the meat of their target" while permitting authors to "retract their errors" by revising their electronic texts. Just as Arthur Kantrowitz imagined the "science court" as a social tool to separate technical judgments from social, economic, and political interests, Drexler saw a future in which writers and researchers used hypertext and Xanadu to "skewer nonsense (such as false limits to growth)," thus "clearing it from the intellectual arena."[57]

Today, of course, this goal appears naïve. The Web is chock-full of erroneous information, some deliberately generated, and one could spend a lifetime trying to fix the manipulated stories presented to the public about highly charged issues like climate change or public health.[58] Meanwhile, maintaining a modicum of civil discourse on the Web is challenging at best. Nonetheless, Drexler's fondness for hypertext, as with his early attraction to Karl Popper's philosophy of science, reflected an optimistic belief of how the scientific community *should* function.

Drexler and Peterson's connections with the Xanadu software project helped foster strong links between the computing community and Foresight. One such bridge was John Walker. A computer programmer and software entrepreneur, Walker cofounded a company called Autodesk in 1982. Its wildly successful computer-aided design packages helped transform the venture into one of the world's top software firms. In 1987, Walker listened to a presentation about Xanadu at a programmers' convention. Soon the "cult hero of Autodesk" took the ambitious project under his wings. The software company supported the promising yet unrealized Xanadu effort with several million dollars before finally cutting its losses and dropping the project in 1992.[59]

In the very same timeframe, Walker also became a proponent of Drexler's nanotechnology. After reading *Engines*, Walker thought that any future in which molecular engineering might become economically important would also give Autodesk an opportunity to "establish itself in the scientific modeling market" and "benefit from the advent of nanotechnology, if and when it emerged." As he understood it, "one could not design atomically-precise structures without a molecular CAD [computer-aided design] system," a potentially profitable product that Autodesk could make and market.[60]

Walker was not the only successful Silicon Valley executive expressing interest in nanotechnology. Mitchell Kapor, a developer of

Lotus 1-2-3 (one of the first so-called "killer apps" for the early personal computer industry) provided early support to Foresight, while smaller donations came from the region's rank-and-file computer and software engineers. "Quite often it would just be some guy who had been a programmer at some company that had a public stock offering. He ended up with two or three million, and he would give a few thousand," Jim Bennett recalled. "That was a typical profile."[61] The Palo Alto chapter of Computer Professionals for Social Responsibility formed a special interest group to consider the future implications of nanotechnology, and Foresight assiduously included computer professionals in the conferences and meetings it organized.[62]

Years later, Christine Peterson offered an explanation for why Drexlerian-style nanotechnology appealed to so many computer scientists and programmers. "They are used to thinking of things digitally, used to taking an engineering approach, used to controlling discrete things," she said. "So they loved the idea that they could do with atoms what you can do with bits."[63] In other words, the ability of computer scientists to control what was happening in their digital world afforded a power that might possibly be translated to the material world. In time, however, other scientists took issue with what they saw as a reductionist, if not sterile, computational approach that neglected working with actual "stuff" and the complexity of chemistry.

Foresight also attracted a disproportionate number of people from the computer science community because of their own field's history. From its birth, the American computer industry had to market itself and the value of its products to skeptical customers. As a result, the industry readily accepted the need for promoting the promise of new technologies. For many researchers, employees, and investors intimately familiar with Silicon Valley's computer industries, Drexler's ideas made sense. Theirs was an environment in which technologies improved at a steady rate. "Having lived through decades of Moore's law–driven miniaturization, with feature sizes going down by orders of magnitude, and seeing the resulting technological effects," computer scientist Mark Miller said "the idea that such progress will continue and be significant is more real to us."[64] Computer science and information technology experts also recognized that, every so often, a new development would appear and shatter the prevailing technological paradigm. Nanotechnology—how it would be achieved, its societal impact, and its inevitability—resonated with a faith in continued technological change and the power of new technologies to remake whole economies.

The New (Nano) World Order

Support and interest from life-extension advocates, computer programmers, and Silicon Valley venture capitalists helped spark broader public interest in nanotechnology. Other, more powerful, historical forces were also in play in the late 1980s and early 1990s that indirectly drove nanotechnology's ascent to "next big thing" status and shaped as its reception among policy makers and business leaders.

One was the continued emergence of the "New Economy." For years, journalists and academics had talked about the replacement of America's traditional smokestack-and-factory manufacturing base with innovative high-technology industries like personal computers, microelectronics, and biotechnology. Knowledge, networks, and innovation, rather than physical resources and heavy manufacturing, seemed to be the new road to jobs, profits, and global markets. As *Time* reported as early as 1983, "Thousands of young people are heeding the cry of the 80's: 'Go West, young man, and grow up with a new industry.' "[65]

At the same time, the Cold War, which fostered the long-term growth of many American industries and corporations, was ending. By 1989, communism was collapsing in Eastern Europe, and the Soviet Union was pulling its troops from Afghanistan. Within a few short years, the Berlin Wall fell, the nuclear arms race ended, and the Soviet Union disintegrated. Political leaders and pundits began to speak of a "new world order" and the "end of history" as liberal democracy and neoliberal economics appeared triumphant.

Yet, as more than four decades worth of bipolarized global structures unraveled, the economic successes of some new "Asian tigers" (e.g., Hong Kong, Singapore, South Korea, Taiwan), along with, especially, Japan's expanding industrial power, challenged American business practices. Corporate executives and economists began to sense a new future full of change, complexity, perhaps even chaos. In fact, chaos and complexity became two of the era's hottest research areas as best-selling science books intimated something radically new was underway in both scientific research and the wider world.[66] The hype that accompanied chaos theory and complexity studies, in fact, was a characteristic that nanotechnology shared. The surge of general interest in all three areas reflected scientists' pursuit of new research topics during a time of global and economic realignment. Meanwhile, media attention highlighting how scientists and economists were constructing theories to explain "nature, human social behavior, life, and the universe itself" helped amplify publicity around these fields.[67]

It would be too simplistic to link the popularity of chaos and complexity studies directly to the changing geopolitical order. Many factors were involved, and the research underlying these fields went back decades. Nonetheless, there were at least indirect connections. For example, throughout the late 1980s, physicists and economists collaborated to study complex systems like global markets.[68] One London reporter, in fact, described chaos studies as a "timely theory which seemed to validate [recent] political turbulence . . . big changes spawned by tiny events."[69]

With so much political and economic uncertainty, corporate executives as well as politicians wanted guidance for dealing with "the postmodern business and consumer" and for anticipating the next major technological trends.[70] Experts predicted that major advances in new areas like nanotechnology would, as with biotechnology and microelectronics, provide a fresh foundation for research, industry, and national economies. "If a revolution of some sort is underway, measuring its impact with any precision has proven to be exasperating," the federal Office of Technology Assessment noted. "Technologies with the power to reshape the basic structure of production have effects where they are least expected."[71] Nanotechnology, some experts imagined, might just be one of these transformative technologies. But how was one to plan for the next technological revolution?

Stewart Brand was one of the well-connected trend shapers who helped business leaders think about and plan for the uncertain future. Just as he had already done for O'Neill and space colonies, Brand skillfully helped raise the profile of both Drexler and nanotechnology for the New Economy's cognoscenti. The most effective tool Brand had at his disposal was the Global Business Network, a small but influential Bay Area–based consulting firm Brand cofounded in 1987. GBN's roots included both 1970s-era futurology and Brand's continued fascination with networks, digital as well as human.[72] As the former Merry Prankster phrased it, GBN offered "survival insurance" to an expanding roster of "companies who want to be smarter."[73]

To do this, GBN constructed "adaptive scenarios ... creative tools for ordering one's perception about possible alternative future environments."[74] It was offering, in essence, an updated mutation of war gaming. During the Cold War, researchers such as Herman Kahn had developed "scenario planning" methods to help military strategists think boldly about "the unthinkable"—that is, play out possible nuclear war situations. Oil giant Royal Dutch Shell concocted similar stories of less controversial "possible futures" to weather the turbulent energy crises of the 1970s. Futurist Peter Schwartz had worked at Shell and known Brand since the early 1970s. After discussing ideas with Jay Ogilvy, a research manager at the Stanford Research Institute, the three men launched GBN.

To some business leaders, GBN's "diverging, plausible, compelling" predictions about the future seemed all the more necessary with the economic downturn and "the emergence everywhere of economic pragmatism."[75] Its corporate clients—these included AT&T, Volvo, as well as numerous energy-related companies—paid upward of $7,000 a day for GBN's services. But how did it create those valuable "diverging" predictions? The scenarios comprising *Limits to Growth*, all of which converged on the same dire outcome, had been generated through a computer modeling of various dire inputs (along with rather trend-conserving assumptions). GBN, by contrast, though it also relied to some extent on current trends projected into the future, also included possibilities for techno-revolution, sociopolitical change, and other new ideas that emerged from its "network" of rather exuberant human minds. The consulting firm boasted an eclectic roster of business executives, artists, academics, and technologists who comprised the firm's expertise. For an annual fee of $25,000, clients could formally join the GBN network and get access to these experts and celebrities. Within three years, GBN was profitable.

Just as GBN's methodology differed from what the Club of Rome had used fifteen years earlier to generate the *Limits to Growth* report, the message had also changed. Whereas the Club of Rome had questioned the continuing quest for economic growth, GBN lauded it and offered upbeat prognostications that emphasized how new markets and new technologies could generate wealth, business opportunities, and positive social change.

But, like the Club of Rome, GBN boasted a loose and yet exclusive membership. A feature article in the techno-hip magazine *Wired* showed some faces from this newly emerging class of high-tech intellectuals: Jon McIntire (Grateful Dead manager), Michael Murphy (cofounder of the Esalen Institute), Gary Snyder (beat generation poet), Bill Joy (founder of Sun Microsystems), Esther Dyson (Freeman Dyson's daughter and technology pundit), Amory Lovins (energy futurist), Jaron Lanier (virtual reality pioneer), William Gibson (cyberpunk author), Laurie Anderson and Brian Eno (avant-garde musicians), and so on. The dotted lines connecting the images of this diverse group suggested an Illuminati-like "collection of heretics and remarkable people."[76] Journalists, of course, had spun similar stories about the Club of Rome's membership in the 1970s. Circa 1990, when stories about impending tectonic shifts in economics and politics were routine, the imagined power of these technologically fluent and geographically mobile elites who could sense the "invisible laws according to which all things functioned" had special salience.[77]

GBN's experts helped shape perceptions of a technological future that, at its boldest, transcended space, bodies, national politics, and perhaps even death. With its "network of remarkable people" (albeit a cohort that one reporter charitably described as "relentlessly white, male, and middle-aged"), GBN provided business leaders with access to new ideas and expertise and created opportunities for the tech-savvy and cosmopolitan elite to meet and mingle with one another at carefully chosen venues.[78] In 1990, for example, GBN members convened at Biosphere 2, a $200 million privately financed experiment outside Tucson based in part on Gerard O'Neill's vision of ecologically self-contained space colonies.

GBN rose to prominence in that brief window of time when the Soviet Union was unraveling and the United States anticipated reaping a "peace dividend." This was, of course, all before Iraq's invasion of Kuwait, the Balkan conflicts, and the emergence of al-Qaeda brought renewed fears and anxieties. At this moment of transition, GBN's evaluation was: "The cold war is over. The Japanese are taking over. Science is sacred. Concern about the environment is a powerful unifying force in the world. We are in an information-based, global economy."[79]

To guide discussions with its corporate clients, GBN prepared three "mental maps of the future" that covered a wide range of possibilities. At one end of the spectrum, "New Empires" predicted protectionist nation-states banding together to create competitive regional clusters and trade barriers. However, if "turbulence and volatility seem to be the only constant," then GBN predicted a second scenario described as "Global Incoherence."[80] Finally, there was "Market World," in which expanding free-market forces and neoliberal economics would drive the creation of a "virtuous circle of technological innovation in an increasingly interactive and prosperous economy."

With Market World's enhanced trade opportunities and favorable geopolitics, GBN and its clients clearly favored the triumph of the third scenario. But, in all the forecasts GBN made, technology figured as one of the prominent forces, if not the most prominent, driving social, economic, and political change. Specifically, GBN imagined that nanotechnology would provide a key set of tools and techniques for the 1990s. How did nanotechnology, still speculative and relatively unknown, emerge as potential prime mover for the coming decades in GBN's outlooks?

Brand and Drexler knew each other, of course, from their space advocacy days. In the mid-1980s, Brand's ties to Drexler were freshened when he visited MIT in advance of writing a best-selling book that profiled how the school's Media Lab was "inventing the future."[81] While studying this much-ballyhooed academic-corporate hybrid, Brand learned about Drexler's ideas for molecular engineering. Brand confessed in *Omni* (and again in his book) that Drexler's nanotechnology left him "giddy, blindsided by a future even more revolutionary than what's coming in computers and communications."[82]

Brand quickly brought Eric Drexler into GBN's orbit. Starting around 1988, Drexler began attending GBN-sponsored retreats and contributing information about nanotechnology that the consulting firm included in its mailings to clients. Brand, who recommended books to GBN's members, also chose *Engines of Creation* as one of his monthly picks. Calling it a "richly prescient harbinger of matters that come to matter a great deal," he advised readers to "hang onto it. . . . [It's] likely to be regarded as historic."[83] Boosted by Brand and GBN, Drexler emerged as a "visionary technologist" with insights for what GBN cofounder Jay Ogilvy called "the post-modern business."[84]

Many Silicon Valley venture capitalists and information technologists were already favorably disposed to Drexler's vision for nanotechnology. GBN helped his ideas reach an even wider, more global audience. At conferences in the United States and overseas, including the World Economic Forum in Davos, Switzerland, Drexler and Peterson spoke about the future of technology. "Five-year projections tell you what your company should be doing now," Drexler told one group in Toronto. "You need twenty-year projections to tell you what your five-year plan should include, and nanotechnology is well within that twenty year span."[85]

Of course, the scenarios Drexler presented at GBN functions were derived from *Engines*, but his appraisals of the technological future were often the first time that business leaders and opinion shapers (who, by and large, were not scientists) learned of the potentially revolutionary technology. GBN's approach to scenarios as a tool for thinking about the future formed the basis for a new book about nanotechnology called *Unbounding the Future*. Written largely by Gayle Pergamit and Christine Peterson, with contributions from Drexler, it presented, à la GBN's method, what Brand called a "rich array of micro-scenarios of nanotechnology . . . some thrilling, some terrifying, all compelling" descriptions of what the (nano)technological future might be like.[86]

Books and scenarios were one way to start a wider audience thinking about nanotechnology. Conferences that brought nano enthusiasts together with businesspeople and scientists offered another approach. In early 1987, Brand read *The Tomorrow Makers*, a pop science book focused on robotics, artificial intelligence, and nanotechnology.[87] In his diary, Brand said that thinking about these radical new technologies "enthuses and scares me," and he imagined holding a conference of techno-future-oriented people to stimulate debate.[88] Two years later, scientists, business leaders, and other nanotechnology enthusiasts met for not one but two conferences to discuss the possibility of manipulating matter at the atomic scale in revolutionary new ways.

In February 1989, Seattle's Nanotechnology Study Group (a self-described " 'Homebrew Club' for Nanotechnology") hosted NANOCON.[89] Neither a "fan gathering . . . nor a true scientific conference," the meeting brought together some eighty "schemers and dreamers" interested in nanotechnology.[90] Greg Bear, who wrote one of the first sci-fi novels to feature nanotechnology, attended as did Gregory Benford, a physicist and sci-fi author from the University of California, and Nadrian Seeman, a scientist from New York University who wanted to use DNA to construct complex three-dimensional nanoscale structures. Surviving transcripts of audience discussions from the NANOCON meeting suggest a small, even intimate, meeting of pro-technology enthusiasts who had gathered to blue-sky about future promises and perils posed by nanotechnology.

The ambience at the first Foresight Conference on Nanotechnology, held in October 1989, was quite different—an international audience, well populated with science reporters, who encountered choreographed and carefully rehearsed presentations. A year earlier, Brand and Peter Schwartz had discussed the absence of any "large conference bringing together all of the players in Nanotechnology."[91] The idea of filling this gap galvanized GBN's leaders, who offered Foresight help so it could stage the event.

From the beginning, Foresight and GBN took pains to make the event as professional and serious as possible. Nano-dazzled undergraduates, for instance, were not invited. A detailed communications and public relations plan, aimed primarily at "scientists and research engineers" along with business leaders and midlevel corporate executives, was prepared.[92] The public relations firm that promoted the meeting saw the event's most important message as "nanotechnology and Eric Drexler are being taken seriously" by industry and academic audiences.[93] The invited audience members represented a diverse range of academic interests: chemistry, physics, mechanical engineering, biochemistry, protein design, and computer science. Stanford, MIT, IBM, Yale, AT&T, Sun Microsystems, and Xerox all sent representatives, as did several venture capital firms. For two days, an impressive roster of scientists, engineers, and business executives discussed whether and when nanotechnology might be the "key manufacturing technology of the 21st century."[94]

With GBN's cosponsorship, the October 1989 meeting generated widespread media coverage as well as a book published by MIT Press. A few weeks after the conference, *Time* ran a story about "incredible shrinking machines" in which microtechnology seemed poised to give way to nanotechnology. Drexler, author of what *Time* called the "nanotechnologist's bible," received especial mention in these treatments.[95] On the other side of Atlantic, the *Economist* published the first of several articles that presented a favorable view of Drexler's ideas.[96] According to *Fortune* magazine, nanotechnology was a major new wealth-generating opportunity for the post–Cold War era, one that bridged the old world of industrial manufacturing with the New Economy's focus on decentralization, flexibility, information, and computation.[97]

Not all coverage was favorable, however, and one can detect as well the first stirrings of the backlash against Drexler and his visioneering that gained force throughout the 1990s. As Gerard O'Neill had already discovered, maintaining command over a popular idea is challenging at best, while controlling how one is presented in the media is often impossible. Part of the difficulty was the tendency of journalists to conflate general predictions about nanotechnology with Drexler's own views. To be sure, journalists often gravitated to Drexler's most expansive claims.

Surprisingly, the *Whole Earth Review*, published by Stewart Brand, presented two of the earliest critiques. Just as he had done for space colonies, Brand wanted to foster debates about new technologies. As a result, Drexler and science writer Simson Garfinkel sparred in a back-and-forth exchange. Garfinkel, a former MIT student, argued that nanotechnology was less a scientific term and more of a "mind set, an ideology, a way of solving big problems by thinking small." Garfinkel charged that "many scientists think Nanotechnology is science fiction." Drexler responded by defending the technical basis of his ideas, challenging his debater to produce "*substantive* criticisms" of his work from scientists. And just as Marx claimed he wasn't a Marxist, Drexler asked that people not instinctively connect all ideas about nanotechnology to him. He stressed that he had not "*advocated* nanotechnology" but rather promoted "*understanding* it."[98] This was by no means the last time he would respond in this fashion to such criticisms.

Steven Levy, author of a best-selling book on computer hackers, seconded Garfinkel's critique in the pages of *Whole Earth Review*. Although he found it "laudable" that the Foresight people had set aside "much time for social and ethical matters," the surfeit of testimonials to Drexler made Levy feel he "was attending Eric's bar mitzvah." Levy suggested that, in the extreme, the "aggregate suspension of disbelief" opened the door to charges of "cultism." Just as serious was the "dissonance" between Drexler's "fully-fleshed vision" and actual lab accomplishments highlighted at the conference, which appeared to Levy as "baby steps towards the moonshot achievement of nanotechnology assemblers."[99]

Perhaps the most serious critique was Levy's observation that, to many people, Drexler *was* nanotechnology. Indeed, there is almost no article about nanotechnology circa 1990 that doesn't quote him, cite his predictions, or offer some interpretation—accurate in some versions and distorted in others—of Drexler's view of the nanotechnological future. This situation was different from what emerged with O'Neill's "humanization of space." Although journalists regularly linked space colonies to O'Neill, the basic idea was already relatively well established, and there was plenty of credit to go around. But when it came to nanotechnology, Drexler literally defined it with his entry in Encyclopedia Britannica's *Yearbook of Science and the Future 1990*.[100] Given the emerging differences between Drexler's supporters and those in the mainstream research community who found his ideas too fanciful—a rift that widened into a chasm over the next decade—one senses that Drexler's visioneering was succeeding perhaps too well.

Computing Chemistry

The early 1990s represent a high-water mark for Drexler's visioneering of a future reshaped by the ability to precisely position reactive molecules and thereby fabricate new devices.[101] For Drexler, the successes of this period occurred at several levels.

For several years, Foresight performed a valuable function for members of the nanotechnology community, especially those in southern California and the Bay Area drawn to Drexler's vision. Memberships in the Foresight Institute climbed after the 1989 conference to around two thousand people.[102] Additional conferences held every few years maintained the trend while a wide range of sponsors, including the Global Business Network, Apple Computer, and Sun Microsystems, helped support these events. Invited high-profile speakers, including previous and future Nobel Prize–winners, attended and presented papers. For example, Rice University chemist Richard Smalley, future Secretary of Energy Steven Chu, Harvard's George Whitesides, and British chemist J. Fraser Stoddart all gave keynote addresses at Foresight gatherings.

Meanwhile, Foresight's newsletter continued to distribute all sorts of news about nanotechnology. Besides noting recent newspaper and magazine articles where nanotechnology (and Drexler) had received attention, the newsletter described recent research developments in areas such as protein design and computational modeling that were relevant to nanotechnology, especially Drexler's biologically inspired vision. Before the World Wide Web facilitated the easy global circulation of information, Foresight also served as a most useful source of news about upcoming nano-related conferences and plans for new research programs in the United States, Europe, and Japan.

In 1991, Foresight spun off a modest-size research initiative, called the Institute for Molecular Manufacturing, that would "advance the technology of molecular manufacturing."[103] With assistance from Ray Alden, a retired telecommunications executive, Foresight raised thousands of dollars for the initiative from Mitch Kapor, John Walker, and others among the Bay Area's high-tech community. Besides helping support Drexler's own research on modeling nanoscale devices, the new institute intended to give seed grants to scientists and produce educational materials.

At about the same time, Foresight also launched a Senior Associates initiative, basing it on a similar program that O'Neill's Space Studies Institute had used.[104] After making a five-year commitment of $250 or more, members of the program, who typically were not scientists but came from a variety of other professional backgrounds, were invited to special events and given opportunities to interact more closely with Drexler. The program worked well enough. For instance, in 1996, Foresight members matched a donation of $40,000 from one senior associate. Besides raising tens of thousands of dollars, these activities helped bring new people into Foresight's orbit.[105]

Finally, in 1993, with encouragement from Richard Feynman's son, Carl Feynman (an MIT alum and Drexler acquaintance), Foresight established a prize named after the famous physicist. The idea of "technological prizes," of course, has a long and distinguished history going back to the time when the British Parliament in the early eighteenth century offered rewards for methods that would allow mariners to determine a ship's longitude. Feynman himself, in his "Plenty of Room" speech, had offered $1,000 to entrepreneurs who could make a supertiny electric motor or miniaturize the information on a book's page to 1/25,000 of its original size. (Less than a year after his talk, an engineer wrote Feynman with proof of his accomplishment and claimed the first prize. The second, to the relief of Feynman, who was paying out of his own pocket, wasn't claimed until 1985).[106]

With efforts led largely by Christine Peterson, the Foresight Institute continued to demonstrate an ability to raise money. Its prize fund, for example, grew such that it could begin giving two biannual awards, $10,000 each, for theoretical and experimental work. In 1996, Foresight announced an even bolder award. A Feynman Grand Prize of $250,000 would go to the first person who successfully performed two tasks: designing and making a "functional nano-scale robotic arm" and building a "digital computing device that fits into a cube no larger than 50 nanometers in any dimension."[107] One of the major donors to the grand prize was Texas businessman James Von Ehr II. The software entrepreneur first encountered Drexler when he heard him give a talk in 1993. Von Ehr was drawn to Drexler's depiction of what nanotechnology could do, but he wanted more details. After reading some of Drexler's more technical publications, he concluded that "the arguments seemed plausible." He later recalled that "as a computer scientist" he could "see the power" of the technologies Drexler described.[108] Von Ehr was also aware of tremendous interest in these technologies among Japanese companies that "were on top of the world in small precise stuff" and sensed a potential business opportunity. After he became involved with Foresight, Von Ehr went on to start Zyvex, a successful Dallas-based company, with the initial goal of building a nanotech "assembler" along the lines of what Drexler advocated in his technical writings.[109]

All these activities—conferences, articles that appeared in peer-reviewed venues, fund-raising, and disseminating news of all sorts related to nanotechnology—positioned Foresight as one of the world's most visible proponents for nanotechnology. Christine Peterson's considerable organization and communication skills helped Foresight maintain a consistent focus on both Drexler and nano-related research and development. As a result, Foresight appeared as a somewhat curious hybrid. It reported on actual research results from labs all around the world but also on ideas for technologies that were very much over the horizon and connected firmly in many observers' eyes to one particular personality. It offered, so to speak, a forum and a place where various communities of futurists, enthusiasts, business entrepreneurs, and researchers from university and corporate labs gathered. But whether they exchanged ideas or simply drifted past one another, literally or intellectually, remained another matter.

As Foresight matured, Drexler continued to receive recognition for his success in promoting nanotechnology. In June 1992, Senator Albert Gore, recently returned from the United Nations Earth Summit in Rio de Janeiro, invited Drexler to testify at a congressional hearing titled "New Technologies for a Sustainable World." At the hearing, Drexler described how "molecule-by-molecule control" could become "the basis of a manufacturing technology that is cleaner and more efficient than anything we know today." Nanotechnology, Drexler told Gore's panel, certainly met "the criteria for an environmentally critical technology," one that could raise the "standard of living worldwide, while decreasing resource consumption and environmental impact."[110]

Drexler's pronouncements resonated with concerns Gore and many in the general public had about the environment and sustainable development as well as improving America's economic competitiveness. Topics that politicians care most about—health, the economy, and national security—all stood to benefit from government investments in nanotechnology. Two weeks after the Senate hearing, Bill Clinton asked Gore to be his running mate in their ultimately successful bid for the White House. While yet a minor blip, nanotechnology was now on the incoming administration's radar screen.

Despite the fact that Drexler had helped bring nanotechnology to the attention of people with real power over policies and purse strings, one charge critics could level against him remained. Unlike most academic or corporate scientists, he had no PhD. In 1991, Drexler remedied this deficiency when MIT conferred a doctorate on him, with Marvin Minsky as his adviser, in the field of "molecular nanotechnology." Drexler encountered roadblocks in his path to *doctor philosophiae* however. As writer Ed Regis describes in his 1995 book *Nano!* Drexler originally planned to get an interdepartmental degree—a fairly normal process for MIT students with academic interests that overlap departments—and have the school's Electrical Engineering and Computer Science Department as his institutional base.[111] The department refused to approve the plan, however. As one faculty member recalled, some people found Drexler's plan of study "flakier than they liked," adding that Drexler didn't want to "go through the standard paths people take."[112] On Minsky's suggestion, Drexler opted to earn his degree through MIT's Media Lab instead.

Despite differing views among MIT's faculty as to the appropriateness of Drexler's dissertation—his degree in molecular nanotechnology was the first of its kind—it served as the basis for his next book. *Nanosystems*, published in 1992 with the subtitle *Molecular Machinery, Manufacturing, and Computation*, was a decided departure from the far-looking *Engines of Creation* as well as from the vague scenarios *Unbounding the Future* presented. Funds from the Institute for Molecular Manufacturing supported research that went into the book's five hundred pages of text, graphs, equations, and figures. Drexler imagined *Nanosystems* as a technical introduction to molecular manufacturing, suitable perhaps for an advanced undergraduate student. By also offering explanations for how the "exploratory engineering" he described in *Engines* might be realized, *Nanosystems* was also an appeal for respectability from mainstream scientists.[113]

In *Nanosystems*, Drexler described, using extensive computer-aided design and simulations, the nano-mechanical gears, bearings, and other components he said could be built. Drexler even offered detailed descriptions of a computer-designed "molecular manipulator." This nanoscale device could, at least in theory, grab, move, and position reactive molecules just as an industrial robot manipulated macroscale objects.[114] Drexler concluded his tome by noting that "advanced molecular manufacturing" might one day be realized by following paths "based on present science but not on present technology." Nonetheless, Drexler included the caveat that his analyses could not yet "be compared to hard-won experi-

Figure 6.2 Publicity photo of Eric Drexler, 1996, holding a model of a nanoscale device (the original caption refers to it as a "nanotechnology robot)." In the background is an image from a computer simulation showing a hypothetical molecular bearing fashioned from a precise number of atoms. (Peter Menzel/Photo Researchers, Inc.)

mental results." Instead they should be seen as an aid "in choosing experimental objectives" and "not as a substitute for achieving them."[115] To some people, this appeared as an invitation. Others saw it as an admission of incompleteness.

Nanosystems received a divided response. The British journal *Nature*'s skeptical appraisal didn't so much question the possibility of what it called "machine-phase chemistry"—building larger molecules and structures by physically bringing atoms and molecules together—as wonder whether this was the optimal path to nanotechnology. With Drexler's approach, "small is not so much beautiful as clumsy."[116] Traditional chemical techniques, *Nature* noted, were an established alternative to futuristic ideas for mechanically positioning individual atoms. For example, scientists had recently built a "molecular train" that could speed around a "chemical track" while pausing at various "stops" along the way.

Molecular marvels like this exemplified a "bottom-up approach to nanotechnology" based on chemistry, not computer simulations.[117] Ultimately *Nature* likened Drexler's computer-based methodology to an "inventor endlessly refining his models" rather than leaping into the messier world of benchtop chemistry for "fear that the real thing will not fly."[118] As some science writers interpreted such skeptical reviews, there was growing opposition to Drexler's ideas among scientists, especially chemists.[119]

Computer scientists could be found on the other side of this nano divide. Drexler's work drew heavily on computer modeling and simulations, and many people with computing backgrounds were intrigued by the idea of making a tangible thing after first designed it in silico with computer code and three-dimensional graphics. At the University of Southern California, for example, a professor in computer science adopted *Nanosystems* for a course named "Molecular Robotics."[120] Reflecting support Drexler found among computer professionals, the Association of American Publishers chose *Nanosystems* as the Best <u>Computer Science</u> Book (as opposed to categorizing it as engineering or chemistry) for 1992.

Ralph Merkle was one computer expert who was enthused about Drexler's vision for nanotechnology as well as the computer-based methodology he employed. Born in 1952, Merkle grew up in the Bay Area and went to Stanford for his PhD. He was no stranger to futuristic technology projects: his father, a physicist at Lawrence Livermore Laboratory, directed Project Pluto for several years, a feasibility study for a device described by one historian as the "nastiest weapon ever conceived." The military envisioned it as a nuclear-powered cruise missile that would roar over treetops at three times the speed of sound, tossing out hydrogen bombs as it went.[121] Merkle's father was responsible for overseeing the design of the reactor that would power the infernal device.

Working with two other Stanford computer scientists, Merkle helped make major advances in the mid-1970s on "public key cryptography."[122] A forerunner to what today is called cybersecurity, this technique allowed for the secure exchange of encrypted information such as digital data over an unsecured communications channel. Now used by computer systems around the world,

Merkle's research helped enable, for example, the secure financial exchanges now at the heart of Internet commerce.[123] Numerous professional societies and the National Inventors Hall of Fame honored him for his contributions.

In 1988 Merkle accepted a job at Xerox's Palo Alto Research Center. Created in 1970, Xerox PARC had a long history of innovative research across a wide range of information technologies. Merkle continued his computer science research but, in the late 1980s, he also became increasingly interested in nanotechnology after hearing Drexler speak at Stanford. Like Drexler, Merkle saw nanotechnology offering a general suite of technological capabilities, especially for medical and life-extension applications. Taking advantage of PARC's considerable computer resources, Merkle began to devote time to what he called "computational nanotechnology," writing more than a dozen articles on this simulation-based approach over the next decade.[124]

Simulations offer researchers another way of producing scientific knowledge, one that complements experimental and theoretical methodologies.[125] Merkle's computational approach to nanotechnology blended two existing research traditions. One came from computational chemistry: starting in the 1950s, researchers used powerful computers to run simulations of chemical reactions and bonding designed from basic scientific principles. Another contribution to computational nanotechnology came in the form of commercial software packages like HyperChem that could design and visualize new chemicals and drugs before they entered production.[126]

Making an analogy to the aerospace industry, which exhaustively designs and "flies" new aircraft on computer screens before even bending a scrap of aluminum, in 1991 Merkle proposed that computers would enable researchers to design and "test" molecular machines. By showing how a particular molecular configuration would, in principle, behave, Merkle said computational nanotechnology could speed the development of real molecular machines by showing researchers the best route to follow before trying to make them in the lab. To show that one could "deliberately design structures ... adequately handled by our computational tools," he described a "molecular bearing" designed with a computer that conformed to basic laws of physics and chemical bonding.[127]

When Drexler had the opportunity to brief Gore's Senate committee the following year, he could show off an even more sophisticated gear system he and Merkle had "built" from 3,557 atoms using advanced off-the-shelf simulation software. Their demonstration was classic visioneering in that it combined technical skills, engineering goals, and a vision of what might be possible in the future. But Drexler acknowledged that, while one could model such molecular devices, the tools needed to build a physical demonstration didn't yet exist. As Drexler told a magazine that covered the Silicon Valley computer industry, computational nanotechnology could let people "see how we can go about developing [the necessary tools] in a step-by-step fashion." Getting the right tools, he acknowledged, however, would still "take many years."[128]

Merkle and Drexler's computation-based studies stirred interest among some scientists and engineers at NASA's Ames Research Center, where O'Neill's space settlement studies had taken place two decades earlier. Besides having easy access to Silicon Valley's computer experts, Ames also hosted NASA's supercomputing center. Originally developed to do simulations of airflow and other fluid dynamics problems, the center had powerful computing resources that could be directed to molecular simulations. Daniel S. Goldin, NASA's head administrator for much of the 1990s, favored bold initiatives, and encouraged the formation of a nanotechnology research group at Ames.[129]

It was a diverse assembly of people and backgrounds. Former L5 Society member Al Globus worked at Ames's supercomputing facility and, inspired by Drexler's oeuvre, imagined that stronger, lighter materials made via nano manufacturing might help make O'Neill-style space settlements possible. Richard Jaffe was an older computational scientist who, before doing simulations of carbon nanotubes, had modeled the effects of space shuttle launches on the chemistry of the upper atmosphere. Another player was Deepak Srivastava, a young theoretical physicist who joined Ames after working at the interface of materials science and chemistry.[130]

The Ames group articulated a number of potential areas where nanotechnology could benefit NASA and space exploration in general.[131] This included, for example, improved data storage, self-replicating machine swarms, and new materials for aerospace applications. Recently discovered molecular forms of carbon, known as buckyballs, fullerenes, and nanotubes, received a great deal of the group's attention. They discussed, for instance, the possibility that one could design and build "atomically precise programmable machines" made of carefully constructed carbon molecules "in the distant future."[132] The group did extensive design and simulation of nanoscale gears and other devices "built and tested" via computer. The group even made special "lenticular images"—the kind where the picture appears to move when you tilt it back and forth—showing their spinning nano gears, and gave these away as promotional items at schools and conference. The Foresight Institute recognized their efforts by awarding Globus, Jaffe, Srivastava, and several other members of the NASA group its Feynman Prize for theoretical work in 1997.

At its peak, some sixty-five people were part of the Ames nanotechnology team, making it one of the world's largest groups focused on nanoscale research. But, as the group continued to expand, its research shifted from computer modeling of nanoscale gears and machines with Ames's supercomputers to the lab-based fabrication of actual materials and devices.[133] "Unless you make something," physicist Srivastava recalled several years later, "people are not going to take it too seriously, right?"[134]

Srivastava's statement captured perfectly the tensions central to the backlash against Drexlerian nanotechnology that started early in the 1990s and continued to grow. While excited about nanoscale research, some academic and corporate researchers found themselves burdened with what they saw as unrealistic expectations for nanotechnology. Antagonism was developing between these researchers and those sympathetic to Drexler's ideas. While some researchers might perhaps attend Foresight events, use Drexler's name to raise awareness about nanoscale research, and even agree with *some* of his ideas, there was a growing sense that little tangible proof of Drexler's concepts existed. As Mark Reed, a Yale physicist and rising star for research on nanoscale electronics, wrote, "There has been no *experimental* verification for any of Drexler's ideas. . . . It's time for the real nanotech to stand up."[135] The question was: when it finally found its feet, which nanotechnology would it be?

CHAPTER 7

Confirmation, Benediction, and Inquisition

||||||||||||||||||||||||||||||||||||||||||||||||||||||||||||||||||||||||||||||||||||||||||||||||||||||||||||||||||||||||||||||||||||||

You and people around you have scared our children.
—Richard E. Smalley, chemist and Nobel laureate,
open letter to K. Eric Drexler, December 2003

Outside the Oval Office, it was a cold, sunny afternoon in early December 2003. Inside, George W. Bush sat behind a desk constructed from timbers salvaged from the nineteenth-century Arctic exploration ship *HMS Resolute*. As cameras clicked, Bush signed the 21st Century Nanotechnology Research and Development Act. It authorized the government to spend nearly $4 billion over the next five years so America would "lead the world at the new frontier of the nanotechnology revolution."[1]

Bush's signature continued the National Nanotechnology Initiative, a program started three years earlier under the Clinton administration. The effort was managed by an ensemble of federal agencies—six at first, but the number expanded to over twenty—and an alphabet soup of committees. Nanotechnology enjoyed broad bipartisan support. By the time Barack Obama sat behind the *Resolute* desk, the U.S. government was spending over $2 billion annually on nanotechnology-related research, making it one of the largest civilian technology development programs enacted during peacetime.

The people invited to stand behind President Bush that December afternoon represented a wide range of interests: a Silicon Valley venture capitalist, a nano-business advocate, a Republican senator, a Nobel Prize–winning scientist. The men all had varying ideas about and expectations for nanotechnology. Mark Modzelewski, from the NanoBusiness Alliance, saw nanotechnology as a potential trillion-dollar business. Chemist Richard E. Smalley envisioned nano as the next major research frontier, just as space exploration had been for him as a teen. Another witness to the signing, Steve Jurvetson, was one of the first venture capitalists to place bets on nanotechnology. As an undergraduate at Stanford, Jurvetson took Drexler's nanotechnology course, and he supported the Foresight Institute and the commercialization of nanotechnology in general.

But Drexler, at times dubbed the "father of nanotechnology," was nowhere in sight.[2] And the bill Bush signed bore scant resemblance to the type of nanotechnology Drexler had long promoted. In fact, by the time the National Nanotechnology Initiative was proceeding full bore, Drexler was "the name that can't be spoken in polite society," or at least among many mainstream scientists and policy makers.[3] How did nanotechnology become a path to the "next industrial revolution" while Drexler, its most visible popularizer and the author of several scientific papers on nanotechnology, find himself so marginalized?

The simple answer: at least two different nanotechnologies emerged in the late 1980s and then matured, sometimes in tandem but often in opposition, through the Clinton era. One was based on Drexler's vision of biomolecular machines "putting every atom in its place."[4] Drexler and like-minded thinkers imagined a technological future that could literally be designed and fabricated from the bottom up. A key methodology was using computer modeling and simulations to predict what could theoretically be built. Drexler's approach reflected Karl Popper's philosophy of science in which the "criterion of the scientific status of a theory is its falsifiability, or refutability, or testability."[5] Even though Drexler claimed his work as more engineering than science, he believed that because no one had demonstrated that molecular assemblers were impossible, they remained a technological path worthy of exploration.

Another nanotechnology that came to the fore in the late 1980s and 1990s was rooted much more in chemistry, physics, and materials science. It was based on the research carried out daily by thousands of academic and industrial scientists and engineers in labs around the planet. Here, work typically progressed incrementally and typified what philosopher Thomas Kuhn called "normal science."[6] Moreover, in Kuhn's classic treatment, scientific discovery was a deeply social process in which dynamics between individual researchers mattered. Whereas Popper described how science *should* work, Kuhn used historical examples to show how it usually *did* work.

As nanotechnology attracted greater media attention and then became a major national initiative, a "crypto-history" of it emerged that camouflaged the contributions from Drexler and others who shared his view of the technological future.[7] Elisions such as this were not uncommon, however. When RCA's David Sarnoff "unveiled" the television in 1939, he gave scant credit to Philo Farnsworth, its key American inventor and promoter.[8] Likewise, in congressional testimonies, annual reports, and promotional materials for the National Nanotechnology Initiative, policy makers and mainstream scientists adhered to a stripped-down "standard narrative." This started with Feynman's insights and jumped ahead to the STM and other Nobel-worthy discoveries without mentioning Drexler.[9]

But an endeavor as sprawling as nanotechnology, which embraces many diverse fields of science and engineering, has many histories. The standard narrative, with its appealing march of progress punctuated by discoveries, is deficient. It hides the essential role played by visioneers like Drexler or Merkle who proffered an expansive view of how molecular engineering, drawing deeply from biology and the physical sciences, could refashion society from the bottom up. Their promotion and proselytizing, and sometimes their research, were instrumental in introducing nanotechnology, or at least a certain version of it, to the broader public.

Throughout the 1990s, at least two nanotechnologies existed in an uneasy yet symbiotic association. In fact, they needed each other. Drexler's version looked to new discoveries made by chemists and physicists—buckyballs, carbon nanotubes, atomic manipulation via STM, nanoelectronics—that could confirm the dream of molecular-scale manufacturing was on its way to realization. Meanwhile, scientists from mainstream research communities cherry-picked futuristic elements from the Drexlerian version to generate increased attention for their work and its potential. Yet, even as they spun their own scenarios of what nanotechnology would enable, they often distanced themselves from the Drexler's visioneering.

By 2003, these two nanotechnologies had diverged from each other in a messy divorce. This left Drexler a pariah among some mainstream researchers who found his popularizing suspect and his ideas questionable. To others, Drexler remained a visionary whose radical idea of "shaping the world, atom by atom"—suggested earlier by people such as Feynman but brought to the public's attention by Drexler and his colleagues—was what had really motivated politicians and businesses to launch lavishly funded nanotech programs.[10] This chapter explores the interdependent relationship between these two technologies, peeling away some of the camouflage that obscures connections, common ground, and conflicts. The first place to start looking is at some of the principal accomplishments that mainstream researchers claimed for the field they increasingly came to call nanotechnology.

Gods of Small Things

His name just invited the puns and bad jokes—Richard E. Smalley. One can imagine headlines science writers concocted before editors ran a line of red ink through them—Small(ey) Science, It's a Smalley World, and so forth. In the early 1990s, Smalley, a mid-career chemist enjoying tremendous success, decided to reframe his entire professional career, putting all other research on hold "until this change to nanotechnology has been made."[11] He eventually became one of nanotech's most articulate and visible advocates.

Born in 1943 in Akron, Ohio, to an upper-middle-class family, Smalley initially had the unlikely ambition to become an opera singer. But Sputnik and the space race launched him into science instead. "Being a scientist or engineer," he recalled, "was one of the most romantic things you could possibly be."[12] His aunt, an organic chemist, nudged him in that direction when he went to college. After getting his undergraduate degree from the University of Michigan, he worked a few years in industry before going to Princeton to pursue a doctoral degree. Smalley's early research focused on cooling simple molecules to slow their rotation and then using laser spectroscopy to study their structure. In 1976, he accepted a position at Rice University in Houston, where he quickly became a rising star.

In 1985, Smalley and several colleagues announced that they had discovered a new allotrope of carbon, a form that was chemically alike yet radically different in its structure from graphite or diamond. After determining that its sixty atoms were arrayed into an almost-spherical cagelike structure, they christened the new molecule "buckminsterfullerene." The molecule's name paid homage to Buckminster Fuller, as the structure of C_{60} resembled the geodesic domes that had made the futurist famous. Researchers excited by the versatile molecules simply called them buckyballs.[13] In December 1991, the journal *Science* honored C_{60} as its "molecule of the year."[14] Five years later, by which time scientists had published hundreds of papers about "Bucky" (Smalley's nickname for the C_{60} molecule), the Rice chemist shared chemistry's Nobel Prize for the discovery.

Smalley's work on buckyballs cemented his reputation, and Rice University administrators worked diligently to retain their star scientist. They gave him power over the hiring of new science faculty and, when the university sought a new provost, Smalley led the search committee. Their choice, physicist Neal F. Lane, later became the head of the National Science Foundation and then Clinton's science adviser. Smalley's ambitions coupled with a charismatic personality—a photo taken shortly before his death in 2005 placed Smalley in a lion tamer's pose before a giant model of the molecule he helped discover—enabled him to construct a well-funded scientific empire.

By the definition I've used throughout this book, Smalley was also a visioneer. Although his professional trajectory was far different from Drexler's or even such as O'Neill, Smalley shared some features. All were graduates of or professors at elite schools, for instance. Especially later in his career, Smalley coupled his scientific expertise to an expansive view of how nanotechnology could be a positive force to remake the future. Although he didn't write books aimed at a broad vernacular public, as O'Neill and Drexler did, Smalley addressed hundreds of audiences from Dubai to Dallas about how nanotechnology could help solve societal problems such as energy shortages and overpopulation as well as poverty and even obesity. Smalley's oft-used catchphrase, now engraved over the entranceway of a building at Rice named after him—"Be a scientist, change the world"—suggests a view toward the future that overlapped with Drexler's even if their approaches to effecting this change differed.[15]

Around 1993, two events happened that shaped the rest of Smalley's career and the history of nanotechnology as well. In June, *Nature* simultaneously published articles by researchers at labs in San Jose, California, and Tokyo announcing the discovery of yet another carbon allotrope. These nanotubes of carbon, only one atom thick, possessed a theoretical strength greater than the finest steel. The discovery electrified the materials research community even more than C_{60} had done.[16] Smalley saw carbon nanotubes as "an incomparable material" with amazing properties, and he plunged into the expanding field of research.[17] He used his clout at Rice to secure a well-funded center for nanotube research and perfected techniques to commercially produce the new forms of carbon.

Smalley also discovered Drexler and the term nanotechnology. While browsing a local bookstore, he found a copy of *Engines of Creation*, which, as he recounted years later, he read "mostly in the bathtub."[18] Although Smalley maintained that Drexler "did something pretty remarkable" in popularizing the idea of nanotechnology, he disagreed with increasing vociferousness as to *how* nanotechnology might be achieved.

After reading *Engines*, however, Smalley realized that the rubric of nanotechnology could provide a powerful strategy for organizing a wide range of research both at Rice and on a wider scale. In January 1993, he briefed Rice's deans about Drexler's nano-assemblers: Although such devices might be "engaging ideas for science fiction, and perhaps not as foolishly naïve as they may sound," they were not something one could use for "choosing research directions over the next decade." However, a robust research program in nanotechnology, which Smalley defined as the "ability to arrange atoms into structures engineered on a nanometer scale," was "reasonable to think about even now." This broader definition embraced, of course, "more than the building of tiny machines." Smalley anticipated that "once our colleagues see the breadth" of opportunities the topic could offer, "their interest may skyrocket."[19]

Smalley pressed Rice's administration to create a Center for Nanoscale Science and Technology that he would direct. In addition to hiring several new faculty, he also insisted on a new campus building. A headline in a Houston newspaper—"New Building to Keep Rice Up with Science and Prof on Campus"—captured the essence of the school's goals.[20] In lobbying for his plan, Smalley made sure that potential donors and campus administrators received copies of Drexler's books as a way of introducing them to the topic.[21]

Smalley, like Drexler, had few qualms about pitching futuristic plans for nanotechnology as he sought resources to enhance Rice's research on carbon nanotubes. For instance, space exploration advocates had discussed hypothetical plans for a "space elevator" for decades. Anchored somewhere in the equatorial regions, the elevator's superstrong yet lightweight cable would reach upward for thousands of miles to a platform in geostationary orbit. Meanwhile climbers, analogous to elevator cars, could ascend the cable powered by solar cells. This technology could allow people and cargo to get into orbit at low cost, thereby revolutionizing space exploration. In 1979, Arthur C. Clarke based his award-winning novel *The Fountains of Paradise* on the construction of a space elevator. In Clarke's story, engineers build the elevator's cable from "pseudo-one-dimensional diamond crystals" he termed "hyperfilament."[22] In reality, the biggest physical drawback to building a space elevator was the lack of a suitable material for the cable.

However, after carbon nanotubes were discovered, Smalley suggested they might be just the right material for a space elevator's backbone. If nothing else, enthusiasm for a space elevator might stimulate lots of new research on carbon nanotubes and bring more attention to the burgeoning new field of nanotechnology. "I have some thoughts that will seem crazy at first blush," he wrote a colleague. "I'm talking with some people at NASA . . . to see what they think about all this fantasizing. Maybe we can suck off a bit of their > $10,000,000,000/yr budget and actually do something worthwhile."[23]

Smalley, if one believes his correspondence and public talks, imagined that a space elevator could actually be built with suitable materials. Of course, it was the space program that had drawn him as a teen to science in the first place. Smalley collected a host of elevator-related studies, corresponded with Arthur C. Clarke, and did some visioneering of his own via detailed calculations as to what properties carbon nanotubes would need to have. He even mailed staff at the White House copies of an article he had coauthored that described a space elevator. Smalley's strategy of blending futuristic and technologically revolutionary possibilities with incremental, day-to-day scientific research worked. By 1999, NASA was supporting his Carbon Nanotechnology Laboratory at Rice with a multimillion-dollar award. Despite the technological promise of carbon nanotubes, Smalley's goal of scaling up their production from pounds to tons per day remained elusive. Moreover, wholesale manufacturing of nanotubes or C_{60} would be more akin to conventional chemical engineering than to nanotech: it would not represent the precise atom-by-atom manipulation that Feynman, Drexler, and even Smalley described.

However, after the scanning tunneling microscope (and its more versatile cousin, the atomic force microscope) was invented, researchers continued to refine the instrument's ability to move atoms around. Drexler even coauthored with an IBM physicist a research letter to *Nature* that described a hypothetical "molecular STM probe tip" that could interact with proteins.[24] Whatever qualms he may have had about the instrument touted so ardently by Conrad Schneiker a decade earlier were gone, and he acknowledged its importance. But these incremental improvements were something only specialists were aware of.

This all changed in 1990 when the *New York Times* announced a nanoscale feat accomplished by scientists working in the rolling foothills above Silicon Valley. At IBM's Almaden lab, Donald M.

Eigler and Erhard Schweizer sprayed a nickel substrate with vaporized xenon, an inert gas. By bringing the probe tip of their STM close to the sample, they learned how to pick up individual atoms and slide them around. In fact, they precisely placed about three dozen of these xenon atoms to spell IBM. The logo they "wrote" was just three billionths of a meter long.[25]

This wasn't the first time researchers had used an STM to move atoms around. But Eigler and Schweizer's work coincided with the media's growing interest in nanotechnology, and this helped ensure that their technical tour de force received widespread attention. The nanoscale IBM logo they made became one of the most iconic scientific images of the 1990s, and other researchers cited their original article in *Nature* hundreds of times. Although his name was almost never mentioned, IBM's atom-moving feat was another confirmation, far more than Drexler's coauthored paper, of what Conrad Schneiker had postulated years earlier. But, as Eigler himself noted, atom manipulation via STM remained a "laboratory tool," not a "manufacturing tool."[26] The carefully prepared crystals on which xenon and other atoms were so delicately placed had to be cooled close to absolute zero, hardly the conditions for assembly-line production. Although the commercialization of STMs and similar instruments was generating healthy sales, neither these instruments nor attempts to commercialize the new carbon molecules seemed poised to radically change American industry for the twenty-first century.

However, the electronics industry was already *incredibly* significant. Worth hundreds of billions of dollars annually, the semiconductor industry of the 1990s, driven by Moore's Law, was fast approaching the point where the dimensions of features on a computer chip would drop below 100 nanometers. Industry experts were eager, if not anxious, to maintain this technological trajectory.[27] Demand for hard disk drives that were both smaller and had greater data storage capacity accompanied consumers' preference for increased computing performance. Given that the already substantial information technologies market was swelling even more with the dot-com bubble, improvements in engineers' ability to pack more data onto smaller disk drives offered the potential for significant profits.[28]

In 1988, a serendipitous discovery paved the way for a marked jump in hard drive performance. Two European scientists, Peter Grünberg of Germany and French physicist Albert Fert, independently discovered that tiny changes in magnetism can produce unexpectedly strong electrical signals in specially prepared materials. The researchers dubbed the new phenomenon "giant magnetoresistance" (GMR), and Fert and Grünberg went on to share the 2007 Nobel Prize in physics for their discovery.[29]

Although GMR was discovered in European labs, it was commercialized in the United States. Engineers initially applied the GMR phenomenon to niche products that required the detection of slight magnetic fields such as land mine–detecting tools and traffic control systems. However, in the 1990s, scientists at IBM's Almaden laboratory, which traditionally focused on data storage technologies, explored its applications for more mainstream products.

Stuart S. P. Parkin was one of the IBM researchers deeply engaged with this research program. After learning about Fert and Grünberg's work, he and his colleagues began to explore the magnetic properties of multilayer films, each a nanometer or so in thickness, with an eye toward improving the capabilities of his company's hard disk drives. As one observer of Parkin's research later recalled, the British scientist and his colleagues "simply engineered the shit" out of the underlying GMR discovery as they made over thirty thousand samples to see how they could apply the nanoscale phenomenon to actual products.[30]

IBM engineers used this research to redesign the "read-write head," a basic element of computer disk drives. Based on the Almaden group's exploitation of the GMR effect, IBM started to manufacture hard drives that were more compact yet had the ability to store a lot more data.[31] IBM made a fortune licensing its technology to other manufacturers, and soon computers around the world had hard drives that incorporated IBM's innovation. IBM's success helped show that devices incorporating new materials and structures, engineered at the nanoscale, could generate a tidy profit.[32] This piqued the interest of policy makers eager to find areas where investments in science had demonstrable economic value.

Buckyballs and carbon nanotubes, the manipulation of atoms with STMs, the exploitation of giant magnetoresistance: these research feats eventually became archetypes for nanotechnology itself. Although just a small selection of nanotechnology's growing research portfolio, they signaled that the label was acquiring greater currency among mainstream scientists.

In 1991, *Science* summarized for its readership the wide range of nanotech-related research underway. Collected in a special section called "Engineering a Small World," one article reviewed chemists' research on "self-assembly," in which molecules spontaneously aggregate to form desired nanoscale structures.[33] Another article described quantized structures such as quantum wells and wires, which confined electron movement to just one or two dimensions, giving rise to new electrical and optical behavior. *Science*'s editors even reprinted Feynman's "Plenty of Room at the Bottom" lecture. Besides reacquainting a new generation of researchers with Feynman's ideas, this also linked nanotechnology to the iconoclastic Nobel laureate. Citations to Feynman's future-oriented talk soon began climbing accordingly.

As more academic and industrial researchers started using the term *nanotechnology*, the question arose as to exactly what nanotechnology was. Did it refer to precise machining and fabrication using techniques that Norio Tanguchi had noted in 1974? Did it mean Feynman's or Drexler's revolutionary nanoscale machines? Or was nanotechnology the building of new materials using techniques that were already familiar to chemists? In many ways, these changing and competing definitions of nanotechnology *are* its history. The ways in which a "new" area of study like nanotechnology becomes a topic that researchers accept as worthy of interest and investigation—its professionalization—provide a wonderful window through which we can see the social life of the scientific community at work. For nanotechnology, a good part of the early activity consisted of researchers actively trying to stake out what the field was and wasn't, and part of this process included defining nanotechnology so that it moved away from the Drexlerian interpretation.

Journals, through whom and what they publish, are key players in demarcating areas of scholarship. In July 1990, the first journal specifically devoted to nanotechnology appeared in researchers' mailboxes. The first editor of *Nanotechnology*, published in the United Kingdom, described the journal's subject as a "new way of thinking" that had first originated "from ultra-precision engineering." But now it epitomized "the drawing together of advanced engineering technology and applied modern physics." At the same time, the journal said it would include the "biological and medical communities" along with scientists doing "engineering, fabrication, optics, electronics, materials sciences." This ambitious goal covered a huge swath of intellectual territory. Moreover, the journal initially described nanotechnology as a "mixture of practice, theory, and computer simulation," which together would develop this "technology of the future."[34] The fact that entire issues of *Nanotechnology* were filled with diverse articles arising from the ongoing series of conferences organized by the Foresight Institute only highlighted the field's catholicity.

Nanotechnology's appearance was just the start of a boom in nano-related publishing. By 2001, some twenty-five new journals related to nanoscale research were circulating.[35] During this efflorescence, scientists, journalists, and policy makers continued their efforts to define and delineate nanotechnology. "Nanotechnology is a very fancy buzzword for chemistry of the next century," one scientist told the *New York Times*. However, the very same article described the manufacture of ever-smaller features on computer chips and also referenced Drexler's exploratory ideas.[36] For other scientists, nanotechnology looked like a "universal discipline" that both straddled and united several different fields.[37] In any case, *nanotechnology* remained a flexible term, one that presented scientists and engineers with new research topics as well as a new administrative category in which to position existing programs and seek additional funding.[38]

One thing was clear though—scientists, reporters, and policy makers referred to nanotechnology more often, to be sure, but in ways that bore little resemblance to the molecular technology Drexler had originally advocated. While Drexler's popularizing helped spur researchers and policy makers to take an interest in nanotechnology, it also set the stage for confrontation as to whose definition would prevail.

Cargo Cult Science?

Starting in the early 1990s, a backlash against Drexler and Drexlerian nanotechnology emerged and grew. When *Science* published a 1991 piece called "The Apostle of Nanotechnology," its title employed a trope frequently used in attacks on Drexler's ideas. Journalists regularly used words like "messiah," "guru," "prophet," and "nanoevangelist" to describe Drexler and, displaying a willingness to span the biblical testaments, critics likened him to both Moses and John the Baptist.[39] Phillip Barth, an engineer at computer giant Hewlett-Packard, took this analogy even further when his posting to the Internet discussion group "sci.nanotech" speculated as to whether "nanoism" might become the "next great mass-movement" in the tradition of Christianity, Islam, or communism.[40] When interviewed for *Science*, Barth dismissed Drexler's visioneering, saying "you might as well call it nanoreligion."[41]

This likening of Drexler to a religious figure was colorful but also ad hominem. Throughout the 1990s, mainstream articles about him included statements from researchers that disparaged him personally. Typically these comments didn't address his research, which had continued to develop after the publication of *Nanosystems*. IBM's Don Eigler told *Technology Review* that Drexler "had no influence on what goes on in nanoscience" and branded his ideas "nanofanciful notions" that might damage the emerging field just as the public's view of it was taking shape.[42] As Drexler saw it, though, many of these scientists lumped fact and fiction together, "used the fiction to discredit the fact," and gave the "resulting stew" the pejorative label "Drexlerian."[43]

Moreover, a few scientists who criticized Drexler admitted they hadn't even read the technical and scientific exposition he gave in his book *Nanosystems*.[44]

To be fair, Drexler was being cut by the other edge of the same journalistic sword that had helped him get such wide attention in the first place. When he rose to prominence in the 1980s, science writers approvingly highlighted his precognition and his eccentricities. Stories about nanotechnology, from *Omni's* 1986 piece onward, had been as much about Drexler as his ideas. Now articles placed some of these same attributes in a less favorable light. But the ad hominem aspects had been there from the start.

Nevertheless, personal attacks are generally uncommon in science as researchers, when quoted for publication, often (but not always) direct their ire toward an idea, not an individual. But, by the mid-1990s, the synecdoche *Drexler = nanotechnology* was established, and attacks on him were, in effect, assaults on his ideas. Drexler responded by pointing out that criticisms of him as a person did not invalidate the principles underlying his exploratory engineering. As he wrote to *Science*, his critics seemed "content to make empty attacks on person and style" while journalists and scientists reduced his central thesis to an "obvious absurdity" and then critiqued it. Provided his critics could "muster a more intelligent argument," Drexler remained ready to defend what he termed the "mechanical control of chemical synthesis."[45]

This perspective—"don't focus on me but on my ideas"—typified Drexler's response to many critics. One might also infer, perhaps, that it reflected Drexler's adherence to a somewhat naïve Popperian ideal as to how science should work. Drexler awaited critics who would falsify or disprove the soundness of his ideas. But, even if the basic science underlying them might withstand scrutiny, some scientists maintained that Drexler's future of molecular manufacturing might simply be too far over the horizon to truly be testable or researchable.

In 1996, *Scientific American*, a magazine widely read by science enthusiasts, made perhaps the most damaging attack on Drexler.[46] A lengthy article by Gary Stix was rife with ad hominem comments that lampooned the "avatar of nanotechnology" and his supporters. How did mainstream scientists and engineers view futuristic concepts for nanotech? "There's no way that one could see of connecting this idea to what we know how to do now," said Harvard chemist George Whitesides, "or can even project in the foreseeable future." Stix's translation? "Come back when you can tell me how to make those things."[47]

According to Stix, nanotechnology à la Drexler was akin to what Richard Feynman had once termed "cargo cult science." In a 1974 Caltech commencement speech, Feynman opined that fads such as UFOs, astrology, and ESP reminded him of Pacific Islanders who, after World War Two, yearned for the return of American troops. To bring the Americans back, they built runways, huts, and radio sets of wood. "They're doing everything right. The form is perfect," Feynman told the graduating students. "But it doesn't work. . . . They're missing something essential."[48] Given that it was Feynman himself who had first presented what many scientists were now seeing as the *ur*-vision for nanotechnology, Stix's comparison was lethal.

Supporters of the Foresight Institute responded vigorously to Stix's article. People such as Ralph Merkle and Jim Von Ehr blasted what they saw as a slanted treatment that was more about Drexler and less about nanotechnology. Even the definition Stix used was wrong, they said. Nanotechnology was not just about making materials and structures "with dimensions that measure up to 100 nanometers," as this would include all sorts of technologies such as genetic engineering and chemical synthesis. No, they said, nanotechnology was "a manufacturing technology able to inexpensively fabricate, with molecular precision, most structures consistent with physical law." Moreover, they claimed the *Scientific American* piece did injustice to Richard Feynman. As the late physicist's son, Carl, told the magazine's editors, "To claim that nanotechnology is cargo cult science because its proponents analyze the capabilities of devices not yet constructed is as absurd as to say that astronautics was cargo cult science before Sputnik."[49]

Besides analogizing Drexler to a religious leader, some people likened his nanotechnology to "pseudosciences" such as alchemy. For example, Phillip Barth reasoned that alchemists had sought the philosopher's stone that might "turn dross into gold," whereas *real* chemists had respectable, fundamental science and new chemical products as their quarry. By the same token, Barth said that "nanoists seek the general-purpose self-replicating nanoassembler to turn dross into diamond," whereas *real* "nanoscientists seek to elucidate basic principles and to develop useful new structures."[50] Although likening Drexler's nanotechnology to alchemy may have scored points with an average reader, historically it was a flawed comparison. As historians of science have shown, alchemy was based on established "scientific" principles—as they were understood at the time—until something like modern chemistry began to emerge in the late seventeenth century and its practitioners rejected earlier practices.[51]

Labeling the work of researchers pursuing Drexlerian goals as "pseudoscience" was inaccurate. Using computer modeling and simulations, for instance, NASA's computational nanotechnologists produced new knowledge that they presented at technical conferences and published in professional journals. Moreover, as a confrontational rhetorical tool, calling something pseudoscience is to try to demarcate "real" research from the "pretend" variety. But it is a fraught category that no one self-identifies with. As historian Michael Gordin has observed, no one goes to their "pseudolaboratory" to do some "pseudoexperiments." Instead, practitioners believe what they are doing is real and legitimate. A category unlike "good science" or even "bad science," the term "pseudoscience" is deployed to demonize—like labeling someone a heretic—when a community feels threatened.[52]

So, if some in the mainstream academic and corporate research communities felt threatened, why might this have been the case? For one thing, scientists have often taken issue with colleagues' popularizing activities, sometimes expressing the view that one should engage the public only at the end of one's research career.[53] For instance, Carl Sagan's "vulgar" works (most notably the television series *Cosmos*, which he did midcareer) supposedly sabotaged his election to the National Academy of Sciences. Gerard O'Neill, meanwhile, became an advocate for space colonies and a public figure only after two decades of work as a respected physicist at an

Ivy League school. Drexler, however, broke from this pattern, publishing his modest oeuvre of "real" research only *after* promoting nanotechnology in a popular book.

Moreover, one of the biggest science controversies of the late twentieth century occurred just as nanotechnology was starting to acquire cachet. In 1989, two chemists claimed they had detected "cold fusion," the production of energy via nuclear reactions at relatively low temperatures. Interest in the work was fueled by journalistic popularization and extravagant claims rather than reproducible, peer-reviewed scientific results.[54] Given that Drexler was promoting nanotechnology with the greatest vigor (and garnering significant media coverage) in the wake of the cold fusion debacle, it is easy to understand how some scientists—especially chemists and physicists—may have been especially wary of popularization and journalistic misrepresentation.

The futurism that pervaded a good deal of Drexler's popular writings and public talks provided another bone of contention. Drexler maintained that the nanoscale devices he spoke of were decades away and, consequently, he was just as likely to be identified as a futurist as a researcher. As he told *Scientific American*, for people who "don't understand that you're talking about the year 2020 or whatever, these ideas raise confused, unrealistic expectations about the short term." This perspective made researchers uncomfortable "because it's not a yardstick they want to be measured by."[55] And while Drexler may have invoked scenarios that smacked of science fiction—connections between Drexlerian nanotechnology with cryonics were especially problematic—the same could be said of researchers like Richard Smalley who spoke at great length about what nanotechnology could or would do. As a continuation of decades of research in chemistry and physics *and* as a major new research frontier, nanotechnology seemed to possess a simultaneous existence in the past, present, and future.

Drexler's own professed ambivalence toward what Thomas Kuhn had famously labeled "normal science"—the incremental research scientists do while working within established paradigms—provided yet another friction point. He took pains to characterize his own work as "exploratory engineering" or, even more abstrusely, "theoretical applied science," which he demarcated from traditional benchtop science. "People in the molecular sciences," he claimed in an interview, "do not know how to think as engineers."[56] When he testified before Al Gore's committee in 1992, Drexler said that expecting chemists and physicists to build molecular manufacturing systems was "like expecting ornithologists to build aircraft." Drexler noted that his own research was done without federal support and suggested that "the need is more for a shift in direction than for a growth in spending."[57] It isn't difficult to imagine a university researcher reading such comments and becoming alarmed or angered.

For all these reasons, two conceptualizations of nanotechnology—Drexler's futuristic visioneering of molecular manufacturing technologies and more incremental research in chemistry, physics, and materials science—collided. The result was Drexler's growing marginalization and diminished ability to effectively promote his visioneering. Reflecting later, Drexler acknowledged the pressure, saying that he "more-or-less exited the field around 1996 hoping to defuse the personalization of it."[58] Whether voluntary or forced, his exit—albeit temporary—from the nano spotlight could have occurred at no more convenient time as scientists, policy makers, and politicians set out to do some visioneering of their own.

Planting the Flag

In 1977, when Drexler first started to think about molecular engineering, the dual revolutions underway in microelectronics and biotechnology shaped his thinking. Twenty years later, federal science managers, politicians, and researchers forged plans for nanotechnology in the midst of the dot-com boom. These circumstances shaped their perspectives about the importance of new technologies and the power of Silicon Valley entrepreneurs to revolutionize economies all around the planet. Now nanotechnology stood poised to be the *new* enabling technology for the early twenty-first century. The title of a report prepared for the cabinet-level National Science and Technology Council—"Leading to the Next Industrial Revolution"—perfectly captured prevailing expectations for economic and social rejuvenation.[59] And, of course, like the dot-com wave, considerable hyperbole preceded the plans for new nano-oriented research programs that started arriving at the desks of policy makers.

Midlevel science managers at various federal agencies played a key role in catalyzing what came to be known as the National Nanotechnology Initiative (NNI). One of these scientist-administrators was James S. Murday. As a young boy, he excelled at science and math and wanted to be "the next Einstein." After training as a solid-state physicist—his 1969 PhD was from Cornell—Murday became the leader of the chemistry research division at the Naval Research Laboratory located just across the Potomac from Washington, DC. Especially interested in surface science, Murday promoted the scanning tunneling microscope as a research tool and organized some of the first nanotechnology conferences. Starting around 1991, he began to tell his Pentagon bosses that the military should be thinking more expansively about new technologies.[60]

One of Murday's counterparts at the National Science Foundation (NSF) was thinking along similar lines. Mihail "Mike" C. Roco was born in Bucharest just after World War Two. After getting an engineering PhD, he emigrated to the United States and, after a series of professorial appointments, became an NSF program director in 1990. A year later, with Roco taking the lead, the NSF started a new initiative to study nanoparticles, bits of nanoscale matter that, owing to their small size, can have properties different from those of larger particles of the same material. Although technically a type of nanotechnology and an important area of materials engineering, nanoparticle research was a far cry from the nanotechnology Drexler (or Feynman) imagined. In 1998, Roco convinced the NSF to launch a new $10 million program for research on what the agency termed "functional nanostructures."[61] The NSF's definition of nanotechnology in the late 1990s was fairly open-ended: "technology that arises from the exploitation of physical, chemical and biological properties of systems that are intermediate in size between isolated atoms/molecules and bulk materials."

Roco, Murday, and a few other science managers began to discuss the possibility of a larger joint effort. Roco recalled that he wanted to promote the "coherence of science and technology" and encourage more cooperation between university and corporate labs.[62] He began to lay the foundation for a much more ambitious program that would involve several federal agencies to support a broad program of research in nanotechnology.

This was a promising time to launch such an entrepreneurial plan. After the Cold War ended, scientists, policy makers, and government agencies debated how to restructure U.S. science policy. National defense against a monolithic communist threat no longer provided sufficient justification for funding basic science. The collapse of the Soviet Union signaled that the next arena of conflict would be in the global marketplace in the form of increased economic competition from European and Asian countries. Articulating an appropriate policy to deal with economic threats from abroad became a priority for lawmakers and policy analysts. Advocates soon touted nanotechnology as an undersupported field with great economic potential for the post–Cold War era.

Toward the end of the 1990s, several prominent people joined Roco's campaign for a national nanotechnology effort. Noticeable strains of techno-utopian visions also started to creep into the speeches of scientists and administrators. For example, physicist Neal Lane said in an oft-quoted statement, "If I were asked for an area of science and engineering that will most likely produce the breakthroughs of tomorrow, I would point to nanoscale science and engineering." It isn't surprising that Lane told legislators about nanotechnology's future applications. After he became Clinton's science adviser in 1998, part of Lane's job was to explain to people on Capitol Hill the down-to-earth benefits that new investments in technologies would bring for health, national security, and the American economy. But Lane's listing of expected benefits also included concepts that, to a layperson, might have sounded similar to Drexler's popularizing, such as nano-size medical probes and "nanomanufactured objects that could change their properties automatically or repair themselves."[63]

Roco also approached Thomas A. Kalil, Clinton's deputy assistant for technology and economic policy. Kalil had previously worked as a trade specialist for Dewey Ballantine, a white-shoe law firm that counted the semiconductor industry among its clients. Kalil understood the needs of companies connected with information technologies and their interest in future technologies. After moving to his White House post, one of major initiatives he worked on was the Next Generation Internet, which aimed to improve business and citizen access to information technologies. As Kalil recalled, after implementation of the Internet initiative, he was seeking another area to focus on when he heard about nanotechnology.[64]

Kalil recognized that nanotechnology, like the Internet, could create jobs and benefit the overall U.S. economy. Roco, for instance, frequently cited studies suggesting nanotechnology could be a trillion-dollar annual market, the realization of which would require some "2 million nanotechnology workers."[65] Kalil also knew that funding for biomedical research had soared throughout the 1990s while support for research in the physical sciences had stagnated. A national nanotechnology program presented an opportunity to redirect money into areas of research that were less well-funded. Finally, Kalil, whose college degree was in political science, saw nanotechnology as the "first critical technology after World War Two where the United States did not start out with a clear advantage." Global leadership in nanotechnology, he recalled, "was up for grabs."[66] As Roco's plans began to take shape, Kalil solicited letters of support from semiconductor manufacturers, the only major industry group that lobbied directly for the National Nanotechnology Initiative.[67]

More generally, nanotechnology's supporters pointed to the success of America's microelectronics and biotechnology sectors as evidence that future-oriented research and development was worth funding. In the late 1990s, as the dot-com boom generated huge economic returns and hyperbolic news reports extolled the benefits of the "information superhighway," this argument was relatively easy to make. In January 1999, Roco invited James Canton, a consultant from the Bay Area–based Institute for Global Futures, to attend a planning meeting for the national nanotechnology program. "Just as the Internet is forcing every business to become an e-business," Canton predicted, "every business in the twenty-first century will become a nano-business."[68] Statements like this suggest how the dot-com era's naïve optimism (what Alan Greenspan famously called "irrational exuberance") provided a model for nanotechnology's future that politicians and policy makers could appreciate.[69]

In March 1999, Roco formally presented his plan for the National Nanotechnology Initiative to Lane and Kalil. At the meeting's end, nanotechnology had two well-placed advocates in the White House. A glossy brochure titled *Nanotechnology: Shaping the World Atom by Atom* made the case for others in the administration and Congress, the majority of whom knew next to nothing about nanotechnology. The colorful publication drew on the nano-utopian rhetoric that was rapidly becoming de rigueur: Feynman's 1959 talk, the unexpected properties of carbon nanotubes, the potential to fabricate nanoscale electronic devices, the now-famous IBM logo writ in xenon atoms, and the vast commercial and societal benefits that investment in the nanotechnology could open up. To evoke the "vastness of nanoscience's potential," the artist who designed the brochure placed an STM-generated image of an atomic "surface-scape" against a background of outer space.[70] The image made a clear association between America's past success in exploring outer space and the new (and waiting) nano frontier.[71] The illustration, however, was also unconsciously ironic, given the interest that Drexler and many of nanotechnology's other early advocates had in the pro-space movement.

In June 1999, a congressional committee called several witnesses to speak of the amazing benefits nanotechnology could offer. Ralph Merkle carried the flag for the computation-driven, Drexlerian view of nanoscale engineering. "Nanotechnology will replace our entire manufacturing base," Merkle said, and "leave no product untouched." This would happen through the development of "artificial, programmable, self-replicating molecular machine systems."[72] In response to a skeptical if somewhat uninformed question from California Democrat Lynn C. Woolsey about the ethics of building such systems—"does self-replicating mean cloning?"—Merkle stressed the health benefits of "surgical tools that are molecular" and the fact that such hypothetical tools would be

Figure 7.1 Cover of the 1999 brochure *Nanotechnology: Shaping the World Atom by Atom.*
(Design by Liz Carroll; image courtesy of the National Science and Technology Council.)

machines, not living creatures reproducing unchecked. He didn't mention cryonics.

Richard Smalley's congressional testimony connected research at the nanoscale with an array of societal benefits including "cancer seeking missiles" and possibly even an "elevator to space." This would be done by building things "on the smallest possible length scales, atom by atom." Smalley shrewdly pointed out his own experience battling cancer and also noted that research in nanotechnology was an "intrinsically small science." Because research would not be centralized in a few national labs, the initiative would spread the federal largesse across the United States and create an "immediacy between the ivory tower pure scientist and the technologist." Funding this "small science" also would attract students into science careers just as the 1960s-era space race had inspired Smalley's generation. All that was needed, Smalley said, invoking both a spirit of conquest and imagery from the first Apollo moon landing, was for someone to "go out and put a flag in the ground and say: Nanotechnology—this is where we are going to go."[73]

Smalley, Merkle, and the others who testified were pushing against an open door. Congressional representatives on both sides of aisle praised nanotechnology's potential rather than probing extravagant claims. Lane and Kalil continued to generate enthusiasm for nanotechnology in the White House. As speechwriters worked on Clinton's final State of the Union address, for example, Kalil recalled that they kept removing references to nanotechnology but the president kept reinserting them.[74]

On January 21, 2000, President Clinton traveled to Pasadena for a major speech on federal technology policy at Caltech's Beckman Auditorium. After an introduction from Caltech alum Gordon Moore, Clinton took the podium before a standing-room-only crowd. An image of the Western Hemisphere constructed by IBM researchers using precisely placed clusters of gold atoms hung above the stage. Clinton warmed up the crowd by joking about how preparing to speak at Caltech had required him to "get in touch" with his "inner nerd." Then the president let his imagination run. What would happen, Clinton asked, if "we could arrange the atoms one by one the way we want them?" Consider the possibilities, he said, of "materials with 10 times the strength of steel and only a fraction of the weight" or "shrinking all the information at the Library of Congress into a device the size of a sugar cube." Similar imagery appeared in his State of the Union address a week later. To make these and other technologies happen, Clinton asked Congress for $3 billion dollars, which would help ignite "the next industrial revolution."[75]

Roco's work had paid off. What had once been perceived as "blue sky research of limited interest," he recalled, or perhaps "even pseudoscience," was now seen as "the key technology of the 21st century."[76] Clinton's benediction contained ideas proposed years, even decades, earlier by forward-looking people such as Feynman and Drexler. But Feynman was dead. And Drexler, like the Moses he was occasionally compared to, was excluded from the land rush of scientists Congress set off by approving the National Nanotechnology Initiative in the fall of 2000. However, the Drexlerian shadow remained over nanotechnology and, over the next few years, it would influence how the American public envisaged the "next industrial revolution" policy makers hoped for.

No Joy in Nanoland

The day after Clinton gave his Caltech speech, an anthropologist interviewed Richard Smalley at his Rice University office. At the end of their conversation, she asked him how he first heard about nanotechnology. Smalley repeated what by then had become a standard story. Around 1992, he read Drexler's *Engines of Creation* and "just thought it was really neat . . . it affected me pretty deeply." But, nonetheless, something about Drexler's vision bothered him. If nano-assemblers and molecular manufacturing were possible, then "all the chemistry we've learned would be irrelevant." In his view, it wasn't possible to just stick molecules together "as if they were Legos." Smalley lamented that too many fantastic images about nanotechnology had already infiltrated popular culture. But, now that the Clinton administration had embraced nanotechnology as a key technology for the twenty-first century, perhaps a different perception of it could emerge. "The word has been cleansed [laughter]," the transcript records Smalley saying. "It's okay now, it's in the mainstream."[77]

What Smalley didn't mention was his own role in helping to remove undesirable Drexlerian strains from nanotechnology. In fact, Smalley made his comments in the midst of an ongoing campaign to promote a form of nanotechnology purged of its Drexlerian past. This process, which was largely complete by the time Bush signed the 21st Century Nanotechnology Research and Development Act, says a good deal about the conflicts between long-term engineering visions and near-term scientific research as well as the place of visioneers within the larger technological ecosystem.

The "cleansing" of nanotechnology seemed essential to people such as Smalley. In the 1970s, Gerard O'Neill's ideas posed no threat to the "space Establishment." They drew little financial support from NASA and produced no major new programs. When it came to moving people off the planet, the space shuttle was the only serious game in town. At worst, O'Neill's ideas were a distraction, perhaps calling attention to NASA's fading glory days of Apollo. At best, his ideas stirred public interest in space exploration.

Nanotechnology presented a different story. By 2001, the National Nanotechnology Initiative was already supporting thousands of scientists and engineers who were working within a growing network of research centers at major universities and national laboratories. Corporations were gearing up to spend more too. Positive public perception was critical if nanotechnology was going to be the major pillar of the "next industrial revolution." The liminal position that Drexler and his visioneering occupied—was it futurism? science? engineering?—posed both a challenge and a threat.

How Americans and other citizens throughout the world thought of nanotechnology was even more critical given the global controversy over genetically modified organisms (GMOs) that had unfolded in the 1990s. Vicki L. Colvin, a leading nano researcher and Smalley's colleague at Rice, told Congress, for example, that "unrealistic scenarios" could thrust nanotechnology into the same "wow to yuck to bankrupt" trajectory that GMOs had experienced.[78]

From a historical perspective, it was a specious analogy. GMOs, and genetic engineering in general, had never enjoyed an unalloyed "wow" period. Criticism accompanied them from the very start. Moreover, organisms—especially those associated with food—*designed* to be released into the environment were almost guaranteed to generate public concern.[79] And, of course, giant agribusiness firms like Monsanto continued to prosper handsomely even after the public's negative reactions to their products. Rhetorically, however, Colvin's analogy was powerful. The idea that the next big technology might indeed follow a "wow to yuck" path alarmed those investing billions of dollars, political capital, or their scientific careers.

Such concerns were not hypothetical, and Smalley's desire to keep nanotechnology "clean" of Drexlerian visions of nano-assemblers became a matter of maintaining a boundary between science and imagined pseudoscience. Starting in the last months of Clinton's presidency and then continuing into the post-9/11 era, several events stirred alarms and necessitated such intervention.

The first shot across the bow of nano advocates appeared even before Congress had approved Clinton's research initiative. The source of the attack was unexpected, as it came not from an environmental activist or techno-Luddite but from "a high priest of silicon."[80] William N. Joy ("Bill" to journalists and colleagues) had founded Sun Microsystems in 1982. Over the next fifteen years, it matured into a Silicon Valley success story worth billions. Joy himself was a Berkeley-trained computer researcher whom *Fortune* once dubbed the "Edison of the internet."[81] But, in the spring of 2000, Joy broke ranks with fellow technologists and wrote an incendiary article titled "Why the Future Doesn't Need Us."[82] Joy's venue for publication—*Wired* magazine—was especially poignant given its trumpeting of the Internet revolution and its cyberlibertarian ideology that deified markets and disparaged regulation. As was the case with Drexler, some journalists depicted Joy via religious analogies such as "the heretic nailing his manifesto to the door of the high-tech temple."[83]

Joy himself had first learned about nanotech through Drexler's writings. Initially entranced by the "utopian future of abundance" that *Engines of Creation* described, Joy later was disturbed by Drexler's warning about the "gray goo" problem—the scenario in which "masses of uncontrolled replicators" proliferate like crabgrass and "obliterate life." Aware of recent controversies over the corporate development of genetically modified crops and the difficulty of containing them, Joy identified self-replication—a key element of the original Drexlerian program—as a clear and future danger. In fact, he estimated the new bio- and nanotechnologies posed a one in three chance of causing human extinction. In response, Joy proposed a radical solution: limiting the development of "technologies that are too dangerous . . . limiting our pursuit of certain kinds of knowledge." [84] In short, he was asking his colleagues to embrace a different "limits to growth" and consciously curtail the expansion of scientific knowledge.

This tocsin might have gone unheard, known only to the high-tech cognoscenti who read *Wired*, but for the flurry of publicity that accompanied it. Major national newspapers interviewed Joy (the *New York Times* ran three pieces about his essay within a week) as did television shows like PBS's *NewsHour*. [85] Although Virginia Postrel, writing for the libertarian magazine *Reason*, blasted Joy for espousing an elitist naiveté that would sacrifice "other people's lives and liberty to his fantasies of power and control," most media attention was sympathetic. [86] The coverage journalists gave Joy's warning especially highlighted the potential perils associated with nanotechnology.

Descriptions of nanobots—a term Drexler himself eschewed—were not new, of course. But stories and pictures of nanobots, particularly "healing machines" patrolling the bloodstream, proliferated as media coverage of nanotechnology spiked following Clinton's speech. The most widely reproduced of these showed a sleek insectoid mechanism plunging a syringe into the red blood cell it gripped. [87] Was the nanobot healing the cell or killing it? The wide distribution of such images certainly had the potential to shape people's outlook for nanotechnology and create unrealistic expectations. James Gimzewski, an IBM researcher renowned for his STM work, recalled his lab getting anxious phone calls after one newspaper announced "IBM creates nanobots that can cure cancer." [88] For ill or good, the hypothetical nanobot had a powerful hold on the public imagination.

Bill Joy's article came at an awkward time for nano advocates like Roco and Smalley as Congress was preparing to vote on Clinton's proposed initiative. Sensing that the question about nanotech's sinister implications was "becoming the determining one," Roco recruited Smalley to help defuse the threat that Joy's article (and Drexlerian nanotechnology) posed to future research and funding. [89] For experienced researchers and science managers such as Smalley and Roco, perhaps Joy's article also brought back un-

Figure 7.2 Artist Coneyl Jay's award-winning rendition of a "nanoprobe" (originally titled *Nanotechnology*) from 2002. It shows a tiny machine giving an injection to a red blood cell. Images such as these were widely reproduced in popular articles about nanotechnology c. 1999–2004. (Coneyl Jay/Photo Researchers, Inc.)

pleasant memories of the late 1960s, when Americans' ambivalence and pessimism about technology posed a formidable obstacle.

In May 2000, Smalley was invited to address a select group of "nationally recognized scientists and innovators" in Washington. "Since Bill Joy's concerns will be a central feature of the discussions," he wrote the meeting's organizer, "I think I will take this opportunity to explain why I don't think we need to worry about self-replicating Drexlerian nanobots." Instead, he wanted to "leave everyone with a notion of the broad nanofrontier that really is in our future."[90] A month later, NASA's head administrator asked Smalley to a private dinner to discuss "impending technological breakthroughs and their effects on the human race." The invita-

tion specifically alluded to "complex and worrisome" ethical issues and a recently emerging "alarmist focus" that accompanied
them.[91] Smalley must have believed his efforts had some effect. "I
hope," he wrote Roco, "that Bill Joy has already hit his high water
mark."[92]

Smalley maintained his efforts to cleanse nanotechnology of its
connections to gray goo and nanobots. A special issue of *Scientific
American* devoted entirely to nanotechnology appeared just before
terrorist attacks struck the United States on September 11, 2001.
Its editors included an article by Smalley called "Of Chemistry,
Love, and Nanobots." It attacked Drexler and Joy head-on by
appealing to constraints on how chemistry worked in real life.
Smalley said that mechanical control of the atoms in a chemical
reaction would require so many "fingers" on a "hypothetical self-
replicating nanobot" that there would be no room for them to
move. Meanwhile, the nanobot's atoms would bond with the very
atoms it was trying to position. Smalley termed these the "fat fingers" and "sticky fingers" problems. Just as no one could make
people fall in love simply by "pushing them together," he continued, a machine that built things "atom-by-atom" might be hypothetically possible, but it couldn't exist in a world where chemical
reactions occur only between "'consenting' molecules." "Such a
nanobot," Smalley assured readers, "will never become more than
a futurist's daydream."[93]

The flames of concern about a negative public reaction to (Drexlerian) nanotechnology gradually died down. But anxieties among
scientists and policy makers were fanned anew in late 2002. On
the Monday before Thanksgiving, HarperCollins published a new
blockbuster by novelist Michael Crichton. Central to the plot of
Prey was the deliberate release of autonomous, self-replicating
nanobots. Created by an amoral corporation working under contract to the Pentagon, this predatory swarm attacked people until
it was destroyed.

HarperCollins carefully choreographed *Prey*'s release to coin-
cide with the holiday weekend. Crichton appeared on talk shows,
gave a book tour, and wrote about nanotechnology for the Sunday
supplement *Parade*, which millions read.[94] After his book became a

number one best seller, rumors even circulated that Hollywood would turn *Prey* into a major motion picture in the same vein as *Jurassic Park*, Crichton's earlier tale of escaped, marauding dinosaurs techno-revived via genetic engineering.

Crichton's new book hit every button that might stoke public alarm about nanotechnology: a greedy, high-tech firm; lack of government regulation; new technologies turned into military applications. *Prey* was, in essence, a fictionalized concoction of everything Bill Joy had warned about. Crichton's book even provided his readers with an epigraph from Drexler, a short introduction to "artificial evolution in the 21st century" (it cited Drexler), and a multipage bibliography that listed Drexler but neglected chemists like Smalley.

Nongovernmental organizations also helped keep controversies about nanotechnology visible for American and European citizens. In January 2003, the Action Group on Erosion, Technology, and Concentration (ETC), an unwieldy name for a small Canadian organization, released a report called *The Big Down*. ETC had previously led campaigns against genetically modified foods. Not surprisingly, its report savaged the idea of nanotechnology. Sensationalistic and sometimes scientifically shaky, ETC's broadside echoed Joy's call for a moratorium on technologies that sought to "manipulate atoms, molecules, and sub-atomic particles to create new particles." Unless regulated, the future masters of what ETC called "Atomtech" would become "the ruling force in the world economy."[95] ETC's scenarios were similar to those the Global Business Network had created years earlier, but this time the corporations were evil, not enlightened. This position reflected ETC's larger interests, which were less about regulating nanotechnology per se and more about restricting corporate power, maintaining cultural diversity, and fostering human rights.

ETC's views got a huge boost when Prince Charles learned of the message in reports like *The Big Down*. The very fact that he was even interested in nanotechnology created headlines around the world ("Charles: 'Grey Goo' Threat to the World"). His publicly expressed concerns about "science that could kill life on earth" seemed to put the Prince of Wales in opposition to Prime

Minister Tony Blair and his plans for revitalizing the British economy.[96] Alarm bells rang again when Astronomer Royal Sir Martin Rees identified nanotechnology as a potential path to extinction and stoked the flames higher by wagering $1,000 that a "bioterror or bioerror" event would cause a million deaths by 2020.[97] Given the global panic after the 2002–3 outbreak of "severe acute respiratory syndrome" (SARS) and the existential fears about terrorists getting weapons of mass destruction, statements such as Rees's were guaranteed to make headlines.[98] Just as Roco's efforts to create the National Nanotechnology Initiative reflected the enthusiasm of the dot-com era, these perceived dangers of nanotechnology mirrored post-9/11 (and, in the United Kingdom, post-7/7) anxieties. As had been the case three decades earlier, when *Limits to Growth* appeared, techno-catastrophism was back.

Saving Nanotechnology

The accumulated animosity between advocates' visions of what nanotechnology *was* and what it *might be* finally burst forth in 2003. As one observer framed it, nanotechnology had finally "reached a crossroads between hypothesis and hype." However, the headlines generated by "one-issue activists, sci-fi thriller authors, demagogic policymakers, and dilettante British royals" were shaping public perceptions.[99]

This conflict between long-term vision and near-term pragmatism came at a delicate time. In 2003, public opinion about nanotechnology was still in its infancy. One survey showed that no more than 20 percent of Americans had even minimal knowledge of what nanotechnology was.[100] A report by the National Research Council warned that speculation about nanotechnology ("a lot of it wildly uninformed") had already "captured the imagination" of the public. Scientists and engineers had already witnessed what this might entail when the "nuclear power and genetically modified foods industries" had ignored the challenges of public perception and "suffered the consequences."[101] Meanwhile, Congress was weighing whether to authorize billions more dollars to fund nanotech research and development. Researchers, science managers, and business leaders were rightly concerned that public disfavor might inhibit their research and derail this investment.

Bothered by the directions this debate was taking, Drexler too became agitated: he wanted to mute the "chorus of false denials and attacks on his reputation."[102] For years, scientists had refused to adequately address the ideas he put forth in *Nanosystems* and other technical publications. Neither the government-funded research nor the monstrous depictions of nanotechnology reflected his visioneering. In the early 1980s, advocates of Reaganesque plans for space-based weaponry had co-opted "the high frontier" from Gerard O'Neill and applied the concept to their own technological plans. Now Drexler sensed something similar happening. "Nanotechnology" was being appropriated and applied to something else while his own work was marginalized, misused, or ignored.

Both Drexler and Smalley, the nano-research community's most visible spokesperson, wanted to save nanotechnology. But which nanotechnology? The answer depended on whom you asked.

The final showdown took place on Smalley's turf: in the pages of *Chemical & Engineering News (C&EN)*.[103] *C&EN's* editor, Rudy Baum, made it the issue's lead story with a cover announcing "Drexler & Smalley Square Off." Even though Smalley's and Drexler's nanotechnologies had co-existed uneasily for years and the two had sometimes borrowed ideas from each other's work, the *C&EN* article set them up in opposition. Ironically, *C&EN's* cover also featured a Drexlerian "molecular mill" showing nanoscale particles being transported and placed "by position rather than reactivity."[104] Inside, through a series of points and counterpoints, *C&EN* used the Drexler-Smalley exchange as an opportunity to simplify competing views of nanotechnology.

It was an odd pairing to anyone not familiar with the years of disagreements over Drexler's visioneering. On one side was an articulate and charismatic person regularly described as a futurist and popularizer, with a relatively small number of technical publications, and no major funding or institutional affiliation. Even Drexler's ties to the Foresight Institute were dissolving following his 2002 divorce from Christine Peterson. On the other side was an articulate and charismatic Nobel laureate from a major university who had secured millions of dollars in grants and authored hundreds of peer-reviewed articles.

Although the *New York Times* reduced the debate to "whether it was possible to build a nanorobot," Smalley and Drexler presented fundamentally different views about what nanoscale research in general could achieve.[105] Despite the seeming opportunity to reach some resolution, Drexler and Smalley largely talked past each other as each stated his own reasons—mechanical versus chemical—why entities such as molecular assemblers could or could not be built. Drexler once again described his ideas as exploratory engineering "akin to pre-*Sputnik* studies of spaceflight" and noted that the very existence of ribosomes, proteins, and cells offered proof of concept. Smalley countered by asking what "new chemistry" Drexler had discovered, since biology could not fabricate "virtually any of the key materials on which modern technology is built." Drexler countered by reiterating how his "machine phase chemistry" could work if researchers just invested enough time in a lengthy "multistage systems engineering effort."[106]

As the debate wound down, Smalley, evincing frustration, claimed Drexler still inhabited a "pretend world where atoms go where you want because your computer program directs them there." His conclusion was an appeal to emotion rather than chemistry, though. Smalley recounted how he had met young students who were "deeply worried," even "scared," about "what would happen in their future as these nanobots spread around the world." Smalley asked Drexler to join him and explain to these children and the public in general that "there will be no monster as the self-replicating mechanical nanobot of your dreams."[107]

Smalley's concluding statement, which conflated Bill Joy's warnings and Drexlerian visioneering, particularly incensed some people. As law professor Lawrence Lessig saw it, Smalley's intent was "stamping out good science" before it had been sufficiently evaluated. "If so-called dangerous nanotech can be relegated to summer sci-fi movies," Lessig opined, "then serious work can continue, supported by billion dollar funding and uninhibited by the idiocy that buries, for example, stem cell research." In other words—give Drexler's ideas a serious hearing and move ahead accordingly. Raymond Kurzweil, a computer scientist and recipient of the National Medal of Technology, opined that Smalley's basic argument—"we don't have 'X' today, therefore 'X' is impossible," as Kurzweil put it—was an unfortunate attitude he had encountered before in his own entrepreneurship.[108] As Drexler had argued for years, Kurzweil said that what living cells do every day was proof that a molecular assembler *could* be built.

With billions of funding dollars and public perceptions at stake, the Drexler-Smalley debate was ultimately about the hegemony of definitions and the marking of boundaries: present-day science versus exploratory engineering; whether the point of view of chemistry or computer science was more representative of the nanoscale world; and, ultimately, who had the authority to speak on behalf of "nanotechnology." Despite his deep roots in the field, Drexler emerged from the debate no less marginalized. Meanwhile, Smalley was frequently invited to reprise his views about what nanotechnology was (and wasn't) until his death in October 2005.

Years later, when reconsidering the attacks he had experienced, Drexler half-jokingly traced his problems back to a misreading of two major philosophers of science. As Drexler put it, he had mistaken Karl Popper's idealized depiction of science as "descriptive when instead he [Popper] was normative" while doing the obverse for Thomas Kuhn, thereby failing to recognize science as the messy social process it often is.[109]

Two days after the "Drexler-Smalley debate" appeared in *C&EN*, Bush signed the 21st Century Nanotechnology Research and Development Act. This occurred amidst rumors that a provision to do a feasibility study of "molecular manufacturing" had been removed at the behest of the NanoBusiness Alliance, an industry association that promoted "small tech."[110] As a sign of how vituperative the debate had become, the alliance's executive director ridiculed the "bloggers, Drexlerians, pseudo-pundits, panderers and other denizens of their mom's basements" for promulgating such an "elaborate fantasy" of backroom political machinations.[111] Nonetheless, an editor at *Nature* noted the bill looked as if it had been written "specifically to avoid mention of the Drexlerian model of nanotech entirely."[112]

Borders, definitions, and debates aside, the first products that explicitly touted their "nano-ness" were already appearing on store shelves. These included sunscreens with nanosize particles of zinc oxide, tennis balls made with "nanoclay," and, most famously, stain-resistant "nano-care" pants. The start of a growing market, yes, but one based on fairly straightforward chemistry, not radical new technologies. Meanwhile, the most important nano-engineered products, in terms of their societal and economic importance—computer chips with ever-tinier features—continued to be produced with little reference as to their "nano-ness."[113] Nanotechnology—a set of possibilities expressed by Richard Feynman, a word coined by Norio Tanguchi, and an ensemble of ideas visioneered by Eric Drexler—had been cleansed.

CHAPTER 8

Visioneering's Value

Grant that he may have power and strength to have victory, and to
triumph, against the devil, the world, and the flesh. Amen.
—Baptism rite, *The Book of Common Prayer and
Administration of Sacraments,* 1827

Over the years, Ed Regis wrote a great many things, including two
books about nanotechnology in general and Eric Drexler specifically. In 2004, the science writer authored one more piece. Compared with the protagonist of his earlier work, the person Regis presented was hardly recognizable. Drexler's "salt-and-pepper beard
and hunched posture" made him "look older than his 49 years."
"Never a rich man," Drexler was now "barely solvent" after his
divorce from Christine Peterson and living in a "modest apart-
ment." He spoke in "apocalyptic terms." And he was angry. "I
never expected that a bunch of researchers would pick up the label
nanotechnology, apply it to themselves, and then try to redefine it,"
Drexler told Regis.[1]

Other publications kept up this theme. After his "debate" with
Smalley, the *Economist* likened "nanotechnology's unhappy father" to an unheeded prophet.[2] And when *Nature* took a retrospective look at the National Nanotechnology Initiative, it didn't
even mention him.[3]

Drexler was down, but he was not out. Regis noted that Drexler
was still "doggedly chasing his dream" of molecular-scale manufacturing. In 2004, he declared that nano manufacturing didn't
necessarily require self-replicating devices. "I wish I had never used
the term 'grey goo,' " Drexler told *Nature* as he backed away from
the most controversial idea associated with him.[4] He also wrote a
counternarrative describing how scientists had co-opted nanotech-

nology to pursue a research program quite different from what Feynman (and he) had originally advocated.[5]

In 2005, the National Academy of Sciences assembled a blue-ribbon panel to evaluate the feasibility of "molecular self-assembly." The presentation Drexler gave to the group compared then-current molecular manufacturing to the state of computing when Charles Babbage built his first calculating machine in the 1820s. Drexler continued to insist that "digitally controlled machines" could "build structures with atomic precision" by following examples biology provided.[6] When the academy released its final report, it concluded that Drexler's ideas still lacked convincing "experimental demonstrations."[7] But it did not judge them impossible. Instead, the report likened Drexler's "visionary engineering analysis" to work done decades earlier by Russian space pioneer Konstantin Tsiolkovskii. Drexler took the comparison as a compliment of sorts.[8] In drawing a connection between Drexler's visioneering and Tsiolkovskii's work, the academy managed to close the circuit between visionary ideas for space exploration and those for nanotechnology. It also kept the door open for future visioneering.

Visioneering's Ripple Effect

Gerard O'Neill's humanization of space and Eric Drexler's atomic-scale manufacturing were both predicated on the idea that the material world could be mastered and controlled. Both men offered cornucopian visions that would transcend pessimistic views found in *The Limits to Growth* and similar reports, and both presented new frontiers for people to explore and exploit. And, of course, the future did not unfold as Drexler or O'Neill had imagined. Nonetheless, their visions captured the public's imagination and stimulated dialogue among and between politicians, scientists, and business leaders. In fact, portraying the visioneering of Drexler and O'Neill as failures because their futures went unrealized underestimates what their visioneering *did* accomplish.

In the case of Drexler, strong evidence of his ideas' influence ironically came from his bête noire, Richard Smalley. The Nobel laureate often credited Drexler for catalyzing the focus of his own research agenda around the rubric of nanotechnology. Although Smalley distrusted the science underpinning Drexler's futuristic scenarios, he recognized the power of these visions to motivate patrons, mobilize other scientists, and galvanize the public. Smalley also believed Drexlerian ideas had a strong potential to spread. The chemist once recalled meeting students who had written essays on the topic "Why I Am a Nanogeek."[9] To his dismay, many students described a future in which Drexlerian nanobots existed and could possibly run amok. Smalley saw this as more evidence that the "darker side of nanotechnology" that people such as Bill Joy and Michael Crichton depicted was finding a toehold in the collective consciousness. These concerns helped stimulate the National Science Foundation to direct millions of dollars toward the study of nanotechnology's societal implications (a tiny fraction of which supported research for this book.)[10]

These examples confirm a larger fact. The advocacy that Drexler, Christine Peterson, and Foresight undertook successfully raised public awareness of nanotechnology. When American politicians supported a federal nanotechnology initiative, it is reasonable to conclude that it was the more radical and transformative Drexlerian vision they were investing in, not the manufacture of stain-resistant pants and skis with carbon nano-fibers in them. Drexler and Foresight did for nanotech in the 1980s what promoters like Walt Disney, Wernher von Braun, and Chesley Bonestell did for space exploration in the 1950s.[11] Their visions shaped expectations, catalyzed people's imagination, and built foundations for later research and development.

To be fair, scientists had been working on the molecular scale long before Drexler. If we imagine the counterfactual case of Drexler's absence, thousands of researchers around the world would still be investigating and building things on the nanoscale. In his absence, we would still have scanning tunneling microscopes, buckyballs, and probably even stain-resistant pants. But without his visioneering, this work might be called something else, and it would probably exist as a much more fragmented research agenda.

Visioneering, as I've defined and deployed it, however, requires more than imaginative and sweeping ideas for how new technologies might dramatically reshape society. It also requires some application of technical skills, knowledge, and calculations to press forward toward the technological future. Engineering, after all, culminates in building *things*. Successful visioneering demands that paper, pencils, and the speaker's podium be exchanged for the hammer, forge, and welder's arc. And this has happened.

For several years, Gerard O' Neill and Henry Kolm supervised MIT students as they built "mass drivers," the electromagnetic catapults O'Neill saw as a first step toward building structures in space. Ted Nelson attracted a cadre of computer programmers to develop software to advance his Xanadu project toward reality. In the 1990s, Drexler, Merkle, and a whole research group at NASA developed elaborate computer simulations of elementary molecular machines. In these cases, visioneering produced more than just detailed designs and blueprints. And, taking a wider view, we can see where others have created new machines, molecules, and institutions that complemented the broader visions people like O'Neill or Drexler imagined for the technological future.

On an early June morning in 2004, for example, an odd-looking aircraft took off from an airstrip in the southern California desert. Attached to the underside of its red and white fuselage was another craft. When the pair reached an altitude of 47,000 feet, the smaller vehicle dropped away, and pilot Michael W. Melvill ignited its single rocket motor. After climbing to an altitude of over sixty miles, hitting Mach 3 speeds along the way, Melvill experienced three minutes of weightlessness before he glided *SpaceShipOne* back to earth. He landed in front of an enthusiastic throng of space buffs. One spectator held up a sign that read "SpaceShipOne— GovernmentZero."[12] Melvill repeated the flight two more times that year, an accomplishment that allowed *SpaceShipOne*'s builders, maverick aircraft designer Burt Rutan and Microsoft billionaire Paul Allen, to claim a $10 million purse from the Los Angeles–based X Prize Foundation. Rutan and Allen later donated *SpaceShipOne* to the Smithsonian Institution, which displayed it

Figure 8.1 *SpaceShipOne* in 2004, after its first flight competing for the $10 million Ansari-X Prize at the Mojave Airport Civilian Aerospace Test Center. Pilot Michael Melvill is shown giving the thumbs-up on touchdown after he flew the space craft to 332,500 feet and landed safely. (© Gene Blevins/Corbis.)

in the dramatic entrance hall of the National Air and Space Museum. It hangs today, appropriately enough, alongside *The Spirit of St. Louis*, which Lindbergh flew across the Atlantic.

As an organization, the X Prize Foundation was the brainchild of entrepreneur Peter Diamandis. In 1980, inspired by O'Neill, Diamandis started a campus group at MIT called Students for the Exploration and Development of Space. Drexler, not surprisingly, attended its first meeting. Riding the tide of early-1980s space enthusiasm, chapters appeared at several other university campuses. After getting degrees from MIT and Harvard, Diamandis eventually launched the X Prize Foundation with Gregg Maryniak, who had worked at O'Neill's Space Studies Institute.[13]

SpaceShipOne sent an unmistakable signal that a new pro-space movement was emerging from the community O'Neill helped inspire. Known variably as "Space 2.0" and "NewSpace," this new cohort of enthusiasts, builders, and entrepreneurs aimed to privatize spaceflight and space launch activities. They also wanted to show that they could put people and cargo into space better and more cheaply than NASA and the world's giant aerospace firms.[14] Here, we find O'Neill's influence, given his belief that space should not be a government-run program but a place. This shift in perspective, so essential to O'Neill's plan for the humanization of space, contributed to the NewSpace movement's foundation.

After *SpaceShipOne*'s flight, other prizes for space-related feats appeared, offering incentives for space enthusiasts seeking the high frontier. The goal of the original X Prize—"To foment human breakout into space. Expand the solution set. End the limits to growth"—would surely have resonated with O'Neill and Drexler, for whom *The Limits to Growth* spurred their visioneering impulse.[15] In fact, the *Limits* thesis, which many economists and other academics had refuted long ago, provided a remarkably durable "windmill" for technological optimists to do battle with.

Like the pro-space movement that O'Neill's visioneering first inspired and the early nanotech community, the NewSpace ethos of the early twenty-first century was shot through with a certain libertarian ideology. It championed free markets, less government, fewer regulations, and well-earned profits for those who pushed the technological envelope. On a photograph of *SpaceShipOne* that Rutan gave to one supporter, he penned "See What Free Men Can Do."[16] Proponents of NewSpace combined a skeptical view of the federal government's ability to foster innovation with a marked enthusiasm for technology's ameliorative economic and social benefits. Unlike the guiding beliefs of the initial cohort drawn to the L5 Society with ideas of space-based settlements as places for social experimentation and communal living, today's NewSpace ethos is relentlessly capitalistic. Many of the investors in new private space ventures made their fortunes in Internet and software companies, people such as Jeff Bezos (amazon.com), Paul Allen (Microsoft), and Elon Musk (PayPal). One longtime enthusiast for private-enterprise space ventures even noted that founding a space-related company had become a "geeky status symbol" for dot-com billionaires looking for something more than a private jet.[17] This new focus departed from O'Neill's visioneering, which initially aimed to benefit society at large, and to meet goals geared more toward personal achievement and benefit.

While their doings were nowhere near as dramatic as a spaceship's launch, academic researchers continued their advance toward being able to build rudimentary nanoscale machines. One line of research, known as "structural DNA nanotechnology," uses DNA itself as a building material. Normally, DNA is a double-stranded linear molecule. Each strand is built from sequences of four chemical bases that assemble into the familiar double helix structure when complementary bases like adenine and thymine join. But, in 1980, crystallographer Nadrian Seeman imagined that it might be possible to synthesize DNA molecules that have single-stranded overhangs called "sticky ends." Seeman proposed starting with carefully designing strands of DNA made with specific sequences of nucleotides. When mixed, these would self-assemble by binding with complementary strands, creating three-dimensional structures.

Over the next thirty years, Seeman worked out the rules that govern how DNA strands can be designed and assembled in order to create more complex shapes and structures. This research was based on the premise that DNA could be a building material and not just a molecule containing genetic information. Seeman's ideas helped foster a growing cadre of researchers who perfected ways to build increasingly complex structures using DNA.[18] Researchers, for instance, managed to turn DNA into a "molecular computer" that could execute mathematical operations.[19] They also used it to build basic "nanomachines." Soon, scientists had fabricated an array of machine-like devices from DNA, including ones that could "walk" along chemical "tracks."

In 2006, Caltech's Paul W. K. Rothemund expanded scientists' capabilities to build nanoscale structures by using "DNA origami." The technique he developed took several hundred short, single strands of DNA that controlled the folding of a single much larger DNA strand, bending it in predictable ways to make complex two-dimensional patterns—triangles, stars, and even smiley faces.[20] *Nature* featured Rothemund's work on its cover, and his accom-

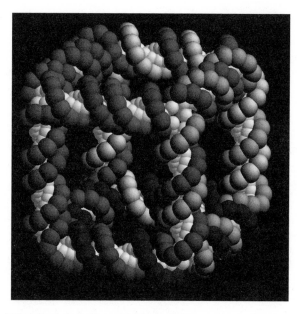

Figure 8.2 A DNA molecule in the form of a cube, potentially a nano-size "building block" that would chemically bond with its neighbors. It consists of six different cyclic strands, each of which corresponds to a face of the cube-like structure. This three-dimensional structure was built using DNA nanotechnology. (Courtesy of Nadrian C. Seeman.)

plishment helped him win a prestigious "genius award" from the MacArthur Foundation. Seeman, meanwhile, shared the 2010 Kavli Prize in Nanoscience for helping develop "unprecedented methods to control matter on the nanoscale."[21]

Soon, major science journals declared that "nanobots," once derided by skeptics, might not be so implausible after all. In 2010, for instance, two different research teams reported that they had extended the capabilities of "DNA walkers." One team managed to get nanoscale "spiders" to follow a trail of nano "breadcrumbs." They started by creating a series of chemical paths on a surface using DNA origami techniques. Then, they built a four-legged walker made of a single-molecule protein as the spider's body and four single-stranded DNA legs. The legs were designed to be flexible, naturally dangling and waving about, and to bind with particular raised points along the paths (the "breadcrumbs") made via the DNA origami techniques. As the legs bent toward the chemically attractive morsels, bonds between the DNA legs and the walking surface broke and re-formed, and the walker automatically moved, spiderlike, along the paths to target locations also chemically "programmed" into the surface.[22] Seeman's team took the "walker" idea one step further. Their DNA machine could pick up gold nanoparticles and carry them, creating, in effect, a "nanoscale assembly line." By demonstrating how it's possible to construct such devices, similar in principle and function to Drexler's molecular assemblers, this research represented modest progress toward building an "autonomous, programmable molecular robot" that "might someday perform tasks too small for humans to do."[23]

Even the phrase "molecular engineering"—introduced by Arthur von Hippel in 1956 and appropriated by Drexler in 1981—acquired new vitality. In March 2011, the University of Chicago announced its new Institute for Molecular Engineering. The school's press office described the multimillion-dollar initiative as deploying "new ways of fabricating and manipulating nanoscale structures to develop new technologies."[24] "This is a singular moment in history," a scientist associated with the endeavor proclaimed, "when our ability to create and control materials at the molecular scale promises to transform the way we engineer solutions to the key scientific and technological challenges of our time." The utopian dream of mastering what J. D. Bernal had once called "the World" remained both durable and desirable.

Transcendence through Technology

Mastery of "the World" notwithstanding, in the era of dot-coms, social networking, electronic commerce, and virtual online worlds, the idea of actually making *stuff*—whether in outer space or in nano space—seemed almost quaint to some people. As one of the cybermanifestos of the early 1990s said, "The central event of the twentieth century is the overthrow of matter. . . . The powers of mind are ascendant over the brute force of things."[25] Although factory workers making semiconductors and iPods would have been surprised to hear about the demise of the material world, statements like this illustrated the continued search for new technological frontiers. They also suggested an opportunity to reconceptualize how existing strands of technological enthusiasm might be woven together anew. For example, in the late 1980s, interest in space, nanotech, artificial intelligence, and life extension all coalesced with the emergence of the quasi-philosophical movement known as "transhumanism."[26]

Short for "transitional human," the word was suggested decades earlier by Julian Huxley, a British evolutionary biologist and brother of *Brave New World*'s author, to reflect what would happen when humanity as a whole decided to "transcend itself" through the "zestful but scientific exploration of possibilities."[27] An essential idea among transhumanists is that new technologies might enable individuals to augment their physical and mental powers and thereby transcend inherent biological limitations. As one early advocate told a journalist in the early 1990s, "I enjoy being human but I am not content."[28] This was a yearning not simply for better technologies but for an improved and enhanced experience of humanity, at least at the individual level.

Despite eugenics' tragic history in the first half of the twentieth century, ideas for technologically improved people percolated among scientists for decades. Even before Yuri Gagarin and Alan Shepard left the earth's atmosphere, for example, medical researchers discussed ways to modify human biology with chemicals and machines to enable long-term space travel, coining the word *cyborg* at the same time.[29] Dandridge Cole and other futurists, as well as scores of science fiction writers, had imagined similar possibilities.

In the fall of 1988, a new zine began to circulate among California's future-leaning techno-hipsters. Two philosophy students at the University of Southern California—Max T. O'Connor and Tom W. Bell or, as they called themselves, "Max More" and "Tom Morrow"—published the first issue of *Extropy*. Their magazine, promoted as a "Vaccine for Future Shock," covered a wide range of technological topics that promised "to radically transform virtually every aspect of our existence." This list, remarkable in its catholicity, included "intelligence-increase technologies, life extension, cryonics and biostasis, nanotechnology." To this, they added space colonization, "economics and politics (especially libertarian)," the "intelligent use of psychochemicals," and "mindfucking."[30] *Extropy* presented an updated and intellectualized version of Timothy Leary's SMI²LE, albeit with the optimistic transcendence-through-technology message.

An even slicker exposition of these ideas, titled *Mondo 2000*, appeared the following year. The semiregular magazine was the brainchild of Ken Goffman (aka R. U. Sirius), a Bay Area denizen fascinated by the "neuro-futurisms" of Timothy Leary and Robert Anton Wilson.[31] Lavishly illustrated, *Mondo 2000* rode the cyberpunk wave started a few years earlier by authors like William Gibson and Bruce Sterling. Continuing where *Omni* stopped, *Mondo 2000*'s audience wanted to know about virtual reality, hacker culture, smart drugs, life extension, and nanotechnologies, all of which were part of *Mondo*'s embrace of what it termed "fringe science."[32]

The *Limits* gauntlet thrown down years earlier by the Club of Rome still provided a potent rallying point. *Mondo 2000*'s debut issue derided the "old" future with its ideals of "going back to the land, growing tubers and soybeans, reading by oil lamps. Finite possibilities and small is beautiful. It was *boring*!" With the Cold War ending and the cyberfrontier beckoning, there was a "new whiff of apocalypticism across the land . . . we are living at a very special juncture in the evolution of the species."[33] This cohort of transhumanists, Extropians, and other techno-enthusiasts often expressed disdain for the material world and even the human body, at least in its original form. A key difference was that a Bernal or a Huxley imagined transformations occurring throughout society or even the entire human species. In contrast, the new transhumanism favored improving the individual via mind and body enhancement (and maintaining a legal right to do so).[34]

By the late 1990s, some transhumanists began to embrace a radical unifying concept called the "Singularity." Its proponents gathered together a wide range of technological ideas—space exploration, nanotechnologies, life extension, artificial intelligence, biological enhancement—into a broader vision for the technological future. Although the Singularity began attracting considerable mainstream attention in the early twenty-first century, it was directly descended from something that had appeared decades earlier in the pages of, not surprisingly, *Omni* magazine. In 1983, *Omni* published a short essay by science fiction writer Vernor Vinge that considered a future in which technological change accelerates at an increasing pace. "When this happens, human history will have reached a kind of singularity," Vinge proposed, "and the world will pass far beyond our understanding."[35]

Debated among technology enthusiasts for several years, the Singularity received considerably more attention after Ray Kurzweil began to promote it.[36] Born in 1948, Kurzweil first achieved national recognition when he appeared as a teen on CBS's show *I've Got a Secret*, where he played a short piece of music. What made the tune special was that a computer Kurzweil had programmed was the composer. Fascinated with broader themes of pattern recognition, Kurzweil experienced early business success while still an undergraduate at MIT by developing software that could match the backgrounds of college students to particular schools. After graduating, he built "reading machines" for the blind that could translate written text into speech, devices that musician Stevie Wonder widely praised and that made Kurzweil a nationally recognized entrepreneur.[37]

Kurzweil also was a staunch supporter of Drexlerian nanotechnology. He participated in events the Foresight Institute organized and vocally advocated cryonics and other life-extension practices. As a futurist, Kurzweil imagined "our biological bodies and brains enhanced with billions of "nanobots," swarms of microscopic robots transporting us in and out of virtual reality.... Human and machine have already begun to meld."[38] Kurzweil wasn't alone in imagining this marriage of silico and vivo. Larry Page, one of Google's cofounders, described a future in which people wouldn't need an Internet search engine. "Eventually you'll have an implant," he mused, "where if you think about a fact, it will just tell you the answer."[39]

Like the pessimistic scenarios in *The Limits to Growth*, Kurzweil's expectations for the Singularity emerged from his basic assumption of exponential growth. Following the example of

Moore's Law, he formulated his own "Law of Accelerating Returns." Kurzweil's maxim posited that the frequency of technological advances in areas such as nanotechnology, artificial intelligence, and biotechnology will increase exponentially until a "rupture in the fabric of human history" occurs.[40] Rooted as it was in selective observations about previous technological trends, critics however saw the Singularity as an "untestable set of assumptions about our near future."[41]

Putting such criticisms aside, in 2009 Kurzweil teamed up with X Prize founder Peter Diamandis to start the Singularity University. Financial backing from Google enabled the venture's founders to organize a curriculum based on a core set of books and ideas that would help students understand the "development of exponentially advancing technologies" and apply these "to address humanity's grand challenges."[42] The topics Kurzweil and Diamandis chose would have been familiar to a reader of *Omni* three decades ago but were now coupled with an eye toward fostering entrepreneurship and social networking: futurology, nanotechnology, artificial intelligence, robotics, space exploration, and neuroscience. Students from around the world competed for spots in Singularity University's summer sessions while CEOs, inventors, and investors plunked down up to $12,000 for week-long "executive programs."

Through their books and blogs, Singulitarians and other transhumanists have engineered a network of supporters, fashioned Internet forums to debate ideas, established institutions, and enrolled philanthropists and patrons to support them. Conferences and summits about the Singularity are annual events that attract hundreds of attendees. Moreover, the location of the Singularity University at NASA's Ames Research Center, where space settlement studies and work on nanotechnology had once been done, suggested a reconvergence of past and present technological enthusiasm.

Just as critics branded O'Neill's space settlements as infantile escapism, critics have likewise attacked transhumanism. "The Singularity is not the great vision for society," one journalist said. "It is rich people building a lifeboat and getting off the ship."[43]

Whereas O'Neill's and Drexler's visioneering suggested a way to escape the limits of an overpopulated planet and reorder the material world, the Singularity is ultimately about avoiding the biological limits of one's own body. Imaginably, these human enhancement technologies could be distributed unevenly throughout society. As a result, political philosopher Francis Fukuyama claimed that human enhancement technologies posed a grave threat to democracy, while conservative bioethicist Leon Kass warned that "human nature itself lies on the operating table."[44] Transhumanism's shades of technological millenarianism—what one critic called "the rapture of the geeks"—are easy to detect regardless of whether technology assumes a transcendent or apocalyptic guise.[45] As one of today's bold statements of the technological future, transhumanism seems poised to inspire future visioneering while fostering debates about its feasibility and desirability.

Assessing Visioneers

Putting philosophy and ethics aside, let's return to the metaphor of a technological ecosystem introduced at this book's beginning. Scholars have long recognized the importance of social, institutional, and economic relations to the success of technological regions. California's Bay Area offers a classic example.[46] Consider the diverse constituents of this particular ecosystem: world-class universities, established manufacturers, ambitious start-ups, entrepreneurs, investors, government funding agencies, the media, along with academic and corporate researchers. All these "species" were present, for example, in the creation of the nanotechnology enterprise in the 1990s, and we can find them in today's NewSpace undertakings as well.

As in a natural ecosystem, there is specialization and diversification in the technological counterpart as well as the exchange of resources such as money, prestige, authority, credibility, and ideas. To mainstream funding agencies, people like O'Neill and Drexler often existed at the margins, but they challenged conventional ideas as to what was technologically possible. Their influence was disproportionate to their numbers and the resources they commanded. And, through their actions and ideas, they helped change the workings of the ecosystem in which they worked. O'Neill, Drexler, and Kurzweil, continuing in the tradition of people like Charles Babbage, Nikola Tesla, and Wernher von Braun, attracted supporters to their particular visions of the future. Meanwhile, the organizations visioneers started or inspired helped spread and amplify their messages throughout the broader ecosystem. Like a real ecosystem, a technological ecosystem with too many visioneers would become unsustainable. Likewise, visioneers cannot survive in all climates. In this story, the connections between Princeton, MIT, and California's Bay Area proved especially conducive to visioneering for space, nanotechnology, and more recently transhumanism. Transplanting an O'Neill or a Foresight Institute to another locale would have produced very different results.

We find both cooperation and competition in all ecosystems. This was true even among those who promoted differing views of the technological future. For example, members of L5 disagreed sharply about the militarization of space, while the sparring between Drexler and Conrad Schneiker can be seen as a struggle for authority and credibility. At the same time, the communities that coalesced around visioneers such as O'Neill shared ideas and fashioned social networks—a process we, in our wired world, take for granted, but a considerable feat in the age of telephone banks and mimeographed newsletters. This helped visioneering ideas to migrate between and among the various small-scale movements devoted to space settlements, molecular engineering, and other exploratory technologies.

Finally, it's impossible to ignore a certain homogeneity among this book's characters. O'Neill, Drexler, Raymond Kurzweil, Peter Diamandis, Ted Nelson, Freeman Dyson, and even Richard Smalley: all are men who graduated from elite schools with technical degrees. To be sure, we find women like Carolyn Meinel, Tasha O'Neill, and Christine Peterson, but they were often organizers and facilitators, not the main standard-bearers for the technological future. Although there is nothing inherently masculine about technology, historically it has been constructed as an endeavor in which men actively participate while women are relegated to the sidelines.[47] Beyond gender, the picture becomes even more homogenous when race is factored in. It is reasonable to ask why visioneering seems an activity primarily attractive or limited to white, privileged males, and it raises questions as to what other narratives of visioneering that go outside this demographic might be like. For example, Asians and Asian-Americans are quite prominent at "geek" high schools like Brooklyn Tech, and at undergrad and grad schools like those of MIT and Caltech they sometimes, or in some programs, outnumber whites. Non-American Asians are obviously prominent in the high-tech world of the Asian Tiger countries. So why do we not see them more often in our cast of visioneers?

Whatever the reasons that compelled them, people such as Drexler and O'Neill, by combining their broad views of the future with technical skills, experience, and research, took speculative ideas out of the hands of sci-fi writers and technological forecasters and put them on firmer ground. Although visioneers' ideas may sit outside the mainstream and require considerable work to establish their legitimacy, their work toward that end secures a beachhead where exploratory notions can exist while entrepreneurial scientists and engineers mobilize and push things one way or the other. By inspiring (or provoking) people, visioneering helps reveal the future as something other than some neutral space that people move into without friction. Instead, it is a terrain made rough by politics, ethics, and economics as well as people's hopes and anxieties.

Power too cheap to meter, the paperless office, hotels on the moon, improved health via genetics, boundless leisure time—such predictions as to what the technological future holds prove remarkable in both durability and banality. Merely demonstrating that something is technologically feasible is no guarantee of success. This was often bitter knowledge for the visioneers. Expansive concepts for the technological future must be matched with equally innovative visions for the social and economic future. Otherwise, we will simply end up with more technologies that fail to meet the futuristic visions accompanying them. Visioneering should not just assume either that today's social or economic inequities will persist or that they will be magically eliminated by technology.

Breaking free of facile scenarios of the technological future is challenging. The roots of technological utopianism go deep in American culture. It is comforting to imagine that a new technology will improve society as promoters of nuclear power, spaceflight, biotechnology, and nanotechnology predicted. Other prognosticators failed to fully appreciate or anticipate major technological changes.[48] The reality of technological change is complex and lies between the poles of revolution and continuity. Moreover, visioneers and their supporters often display a disconcerting tendency to fall in love with their ideas, promoting them in the face of balanced criticisms. As a result, visioneering can display a dark side that is neither healthy for the technological ecosystem nor helpful to the general polity. As with technology itself, when asking about visioneering's value, perhaps it's best to begin with the obvious question "For whom?" It would be a mistake to conclude that visioneering is always benign.

Visioneering Tomorrow

A person who fell asleep in 1972 with *The Limits to Growth* on their lap and then woke up today would find today's headlines eerily familiar. As I write this, Jerry Brown is again governor of California while newspapers announce soaring oil prices, shortages of key minerals, and the birth of Spaceship Earth's seven billionth inhabitant. In response to dire expectations, "doomsters" have suggested the need for restraints and restrictions, ideas that their "boomster" counterparts resist.[49] Language redolent of the *Limits* era is circulating once again. When NASA ended its space shuttle program in 2011, one writer said it symbolized "a time when limits are all we talk about. Today we have no stars in our eyes."[50] Meanwhile, another expert noted that "we are now using so many resources that we have reached some kind of limit" unless we see major technological improvements.[51] Warnings of limits provided a powerful motivation for visioneers like Drexler and O'Neill and will doubtless encourage new people to propose paths for technological futures.

Consider the field of synthetic biology. Its engineering-oriented proponents imagine using biological "parts" such as standardized DNA sequences (sometimes called BioBricks™) to build new organic entities just as an engineer might construct a circuit from basic electronic components. In journalists' accounts, practitioners of "syn bio" will be able to "re-design the living world" and create "tiny, self-contained factories" that make "cheap drugs, clean fuels, and new organisms."[52] In fact, Drew Endy, the field's "most fervent evangelist," called synthetic biology a "nanotechnology that actually works."[53] Like Drexler, Endy trained in engineering and worked at MIT's Computer Science and Artificial Intelligence Laboratory. NASA's Ames Research Center and the Defense Advanced Research Projects Agency (DARPA) have launched synthetic biology initiatives.[54] Meanwhile, venture capitalists are expressing interest in synthetic biology's commercial possibilities while groups like ETC have prepared reports opposing "extreme genetic engineering" and "syndustry."[55] In almost every way one looks at it, synthetic biology's recent history is strikingly similar to the visioneering for nanotechnology.[56]

In a 2009 book titled *Whole Earth Discipline*, Stewart Brand championed a range of exploratory engineering ideas as part of his "ecopragmatist manifesto." Riffing the famous maxim he used to launch the *Whole Earth Catalog* in 1968, Brand proclaimed, "We are as gods and HAVE to get good at it."[57] Synthetic biology was one of the new godlike powers Brand suggested humanity might need in order to survive. Geoengineering, the deliberate modification of the planet's climate as a solution to global warming, was another.[58] Both technologies represent the resilient goal of mastering the material world. And, like so many of the visioneers he interacted with over the years, Brand was not terrestrially bounded. In 2011, the *New York Times* announced that DARPA would sponsor a public symposium to jump-start a "persistent, long-term, private-sector investment . . . to make long-distance space travel practicable and feasible."[59] Visioneers of all stripes appeared on the symposium's program as did Brand, who chaired a panel that considered the societal implications of interstellar travel. The meeting's organizers looked to a "technological horizon . . . replete with interesting problems," presumably ones similar to those that had captivated people such as O'Neill and Drexler.[60]

While ideas for an interstellar starship make for great news copy, futuristic technological schemes such as geoengineering appear as attempts to evade hard political, social, and economic choices. If scientists and policy makers are going to seriously consider such technological solutions for solving contemporary crises, knowing the histories of unrealized technological futures is essential. Technological futures come with social and economic consequences that we must take into account. Technology has indeed given us godlike powers, but they are ones we have yet to master.

The challenge of overcoming and circumventing limits motivated the visioneers described in this book. Tomorrow we will face new limits. Thinking in measured and mature terms about our technological future is one way to confront and perhaps overcome them. Decades ago, J. D. Bernal began his book *The World, the Flesh, and the Devil* with an evocative and eloquent statement: "There are two futures, the future of desire and the future of fate, and man's reason has never learnt to separate them." Technologies are ultimately tools we use to consciously construct our future rather than simply accepting fate. Visions of the technological future have helped catalyze action and innovation. The choice between the future we want and the one we ultimately make is ours.

A Note on Sources

||||||||||||||||||||||||||||||||||||||||||||||||||||||||||||||||||||||||||||||||||||||||||||||||||||||||||||||||||||||||||||||||||||||||||

Writing a history of technological futures presents challenges as well as opportunities. Because it is contemporary history, many of the traditional sources a historian relies on have not passed through a professional archivist's hands and entered formal collections. The story I tell here is necessarily incomplete. Much of the evidence on which this book is based came from people's personal papers and, as a result, has never been examined systematically by a historian before. I found these materials in basements, garages, and, in one memorable case, a storage unit near the San Jose airport with a remarkably resilient lock.

Writers and reporters often provide a "first draft" of history. Science journalists' contemporary depictions of space settlements or nascent nanotechnology are vivid and often designed to generate wonder or bemusement. Their stories about the technological future naturally generated public interest, and the more spectacular, it seems, the better. They convey an immediacy, to be sure, but often lack a critical perspective. However, as this book's aim is to explain how people thought about the technological future at different points in recent history—from the mid-1970s to the end of the Cold War to the post-9/11 era—these journalistic accounts offer insights into how people encountered radical ideas for technological futures at particular points in time.

Interviews proved another valuable resource. Most of the interviews I used were conducted by myself either in person or by telephone. A few longer interviews were formally transcribed and copies placed either with the Chemical Heritage Foundation in Philadelphia or the Center for History of Physics in College Park,

Maryland. In all cases, interviewees gave written or oral consent to be interviewed for my research. In a few cases, I used interviews and oral histories done by colleagues. These are indicated along with where copies can be found. Throughout my research and writing, I tempered the use of interviews with the understanding that personal memories are selective and change over time.

Yet despite all the sources mentioned above, documents—both privately held materials and those that have made their way to formal archives and the published record—provided the main evidentiary foundation on which I built my narrative. The archival and primary source materials cited in this book's notes are identified with the following abbreviations:

ARK/DC   Papers of Arthur R. Kantrowitz (MS-1097), Rauner Special Collections, courtesy of Dartmouth College Library, Hanover, NH.

CHF   Donald F. and Mildred Topp Othmer Library of Chemical History, Chemical Heritage Foundation, Philadelphia, PA.

CWS   Conrad W. Schneiker personal papers; copies in author's working papers.

FD   Freeman Dyson personal papers; copies in author's working papers.

FI   Foresight Institute working files; copies in author's working papers.

GEO/NASM   Geostar Corporation Records, 1983–91, Accession number 2006-0049, National Air and Space Museum Archives, Smithsonian Institution, Washington, DC.

GKON   Gerard O'Neill personal papers, currently maintained by Tasha O'Neill in Princeton, NJ; copies of material cited in author's working papers.

GKON/PUA   Gerard O'Neill papers, Faculty File and Communication Office Records (box 66, folder 12, and fox 67, folder 1), Seeley G. Mudd Manuscript Library, Princeton University Archives, Princeton, NJ.

HK   Henry Kolm personal papers; copies in author's working papers.

JCB   James C. Bennett personal papers; copies in author's working papers.

KH/SA Papers relating to Project Xanadu, XOC, Memex, Ted Nelson, and Eric Drexler, 1977–97 (M1292), Stanford Special Collections at Stanford University, Stanford, CA.

KH-AL/SA Materials donated by Keith Henson and Arel Lucas to Stanford University Archives, unprocessed when examined July 2008, now listed as M1291, Stanford Special Collections at Stanford University, Stanford, CA.

LTG/SIA Papers from 1972 Limits to Growth Conference, RU 275, Smithsonian Institution Archives, Washington, DC.

NASA/HO Papers from NASA History Office, NASA Headquarters, Washington, DC.

PF/SA Point Foundation records (M1441), Stanford Special Collections at Stanford University, Stanford, CA.

RAH/UCSC Robert A. Heinlein papers at the University of California, Santa Cruz, CA.

RS/CHF Richard Smalley Papers, Donald F. and Mildred Topp Othmer Library of Chemical History, Chemical Heritage Foundation, Philadelphia, PA.

RS/RU Richard Smalley Papers, Rice University, Houston, TX.

SB/SA Stewart Brand papers (M1237), Stanford Special Collections at Stanford University, Stanford, CA.

SSI Material from the Space Studies Institute files, originally in Princeton, NJ; copies in author's working papers.

WEC/SA Whole Earth Catalog Records (M1045), Stanford Special Collections at Stanford University, Stanford, CA.

Copies of all documents I cite are in my personal working files. Following the publication of this book, these will be offered to Stanford University's Department of Special Collections.

Notes

||||||||||||||||||||||||||||||||||||||||||||||||||||||||||||||||||||||||||||||||||||||||||||||||||||||||||||||||||||||||||||||||||||||

Introduction: Visioneering Technological Futures

1. O'Neill's term appears in a number of articles by and about him throughout the 1970s such as Richard K. Rein, "Maybe We Are Alone," *People*, December 12, 1977, 123–35.

2. Gerard K. O'Neill, *The High Frontier: Human Colonies in Space* (New York: William Morrow and Company, 1977).

3. Michael McClure, *Antechamber and Other Poems* (New York: New Directions, 1978), 43.

4. For one description of Space Day, see Roger Rapoport, *California Dreaming: The Political Odyssey of Pat & Jerry Brown* (Berkeley, CA: Nolo Press, 1982), 186–92.

5. George Koopman, "We're Going," *L5 News*, September 1977, 2–4.

6. Ibid.

7. R. Crumb, "Space Day Symposium," *CoEvolution Quarterly*, Fall 1977, 48–51.

8. Rapoport, *California Dreaming*, xiv.

9. Ibid., 183–92.

10. From Brown's August 11, 1977, speech, folder 10, box 10, WEC/SA.

11. William Endicott, "Brown, on TV, Talks Issues," *Los Angeles Times*, June 26, 1976, A5.

12. John C. Bollens and G. Robert Williams, *Jerry Brown in a Plain Brown Wrapper* (Pacific Palisades, CA: Palisades Publishers, 1978), 17.

13. Lizabeth Cohen, *A Consumers' Republic: The Politics of Mass Consumption in Postwar America* (New York: Vintage Books, 2003).

14. Descriptions taken from various sources including Robert Gillette, "The Limits to Growth: Hard Sell for a Computer View of Doomsday," *Science* 1972, 175, 4026: 1088–92; Claire Sterling, "Club of Rome Tackles the Planet's 'Problematique,'" *Washington Post*, March 2, 1972, A18; David Anderson, "A Careful Look at Growth as Suicide," *Wall Street Journal*, March 17, 1972, 6; Peter Passell, Marc Roberts, and Leonard Ross, "Review: The Limits to Growth," *New York Times*, April 2, 1972, BR1; Bowen Northrup, "Thinking Big: Club of Rome Merges World's Woes into One," *Wall Street Journal*, October 2, 1972, 1; and Samuel C. Florman, "Another Utopia Gone," *Harper's*, August 1976, 29–36.

15. Donella H. Meadows, Dennis L. Meadows, Jørgen Randers, and William W. Behrens, *The Limits to Growth* (New York: Universe Books, 1972); hereafter, *Limits*.

16. Walter Sullivan, "Struggling against the Doomsday Timetable," *New York Times*, June 11, 1972, E7.

17. K. Eric Drexler, "Mightier Machines from Tiny Atoms May Someday Grow," *Smithsonian*, November 1982, 145–54.

18. K. Eric Drexler, *Engines of Creation: The Coming Era of Nanotechnology* (New York: Anchor Books, 1986).

19. I claim no credit for coining the word *visioneer*. A quick Internet search reveals, among various things, its use for a Christian self-help book, an independent film, several companies, and trademarked commercial products. However, I'm the first to suggest employing *visioneer* as a term of historical analysis and category of historical actors, which is how it is used here.

20. David E. H. Jones, "Technical Boundless Optimism," *Nature* 1995, 374, 6525: 835–37.

21. John Law, "Technology and Heterogeneous Engineering: The Case of Portuguese Expansion," in *The Social Construction of Technological Systems: New Directions in the Sociology and History of Technology*, edited by Wiebe Bijker, Thomas Hughes, and Trevor Pinch (Cambridge, MA: MIT Press, 1987), 111–34.

22. Frank Winter, *Prelude to the Space Age: The Rocket Societies* (Washington, DC: Smithsonian Institution Press, 1983).

23. Howard E. McCurdy, *Space and the American Imagination* (Washington, DC: Smithsonian Institution Press, 1997); Michael J. Neufeld, *Von Braun: Dreamer of Space, Engineer of War* (New York: Alfred A. Knopf, 2007).

24. For example, Martin Kenney, ed., *Understanding Silicon Valley: Anatomy of an Entrepreneurial Region* (Palo Alto, CA: Stanford University Press, 2000); Judy Estrin, *Closing the Innovation Gap: Reigniting the Spark of Creativity in a Global World* (New York: McGraw Hill, 2009).

25. John D. Douglas, "The Future of Futurism: An Analysis," *Science News* 1975, 107, 26: 416–17.

26. Edward Cornish, "Future Shock and the Magic of the Future," *Futurist*, November–December 2007, 43–46.

27. William H. Honan, "The Futurists Take over the Jules Verne Business," *New York Times*, April 9, 1967, M243.

28. Jennifer S. Light, *From Warfare to Welfare: Defense Intellectuals and Urban Problems in Cold War America* (Baltimore: The Johns Hopkins University Press, 2003).

29. H. Wentworth Eldredge, "The Mark III Survey of University-Level Futures Courses," in *The Future as an Academic Discipline* (New York: Elsevier, 1975), 5–18.

30. Jib Fowles, "Review: The Study of the Future," *Technology and Culture* 1978, 5, 29: 34–35; David E. Nye, *America as Second Creation: Technology and Narratives of New Beginnings* (Cambridge, MA: MIT Press, 2004).

31. William A. Sherden, *The Fortune Sellers: The Big Business of Buying and Selling Predictions* (New York: John Wiley & Sons, 1998); David Rejeski and

Robert L. Olson. "Has Futurism Failed?" *Wilson Quarterly*, Winter 2006, 14–21; Matthew Connelly, "Future Shock: The End of the World as They Knew It," in *The Shock of the Global: The 1970s in Perspective*, edited by Niall Ferguson et al. (Cambridge, MA: Belknap Press of Harvard University Press, 2010), 337–50.

32. For a critique, see chapter 12 of Howard P. Segal, *Future Imperfect: The Mixed Blessings of Technology in America* (Amherst, MA: University of Massachusetts Press, 1994).

33. Nik Brown et al., *Contested Futures: A Sociology of Prospective Techno-Science* (Aldershot, UK: Ashgate, 2000); Marita Sturken, Douglas Thomas, and Sandra J. Ball-Rokeach, eds., *Technological Visions: The Hopes and Fears That Shape New Technologies* (Philadelphia: Temple University Press, 2004).

34. Matthew N. Eisler, *Overpotential: Fuel Cells, Futurism, and the Making of a Power Panacea* (New Brunswick, NJ: Rutgers University Press, 2012).

35. From Jim McClellan, "The Tomorrow People," *UK Observer*, March 26, 1995, reprinted in *Ottawa Citizen*, April 1, 1995, B4.

36. Howard P. Segal, *Technological Utopianism in American Culture* (Chicago: University of Chicago Press, 1985),; Joseph Corn, ed., *Imagining Tomorrow: History, Technology, and the American Future* (Cambridge, MA: MIT Press, 1986); also, Jay Winter, *Dreams of Peace and Freedom: Utopian Moments on the Twentieth Century* (New Haven, CT: Yale University Press, 2008), and essays in Michael D. Gordin, Helen Tilley, and Gyan Prakash, eds., *Utopia/Dystopia: Conditions of Historical Possibility* (Princeton, NJ: Princeton University Press, 2010).

Chapter 1: Utopia or Oblivion for Spaceship Earth?

1. "Text of President Johnson's Speech at Dedication," *New York Times*, April 23, 1964, 26.

2. See chapter 9 of Thomas P. Hughes, *American Genesis: A Century of Invention and Technological Enthusiasm* (New York: Penguin Books, 1989), as well as Everett Mendelsohn, "The Politics of Pessimism: Science and Technology Circa 1968," in *Technology, Pessimism, and Postmodernism*, edited by Yaron Ezrahi, Everett Mendelsohn, and Howard Segal (Dordrecht: Kluwer Academic Publishers, 1994), 151–74, and Timothy Moy, "The End of Enthusiasm: Science and Technology," in *The Columbia Guide to America in the 1960s*, edited by Beth Bailey and David Farber (New York: Columbia University Press, 2001), 305–11.

3. Lizabeth Cohen, *A Consumers' Republic: The Politics of Mass Consumption in Postwar America* (New York: Vintage Books, 2003).

4. David Zierler, *The Invention of Ecocide: Agent Orange, Vietnam, and the Scientists Who Changed the Way We Think about the Environment* (Athens: University of Georgia Press, 2011); Adam Rome, " 'Give Earth a Chance': The Environmental Movement and the Sixties," *Journal of American History* 2003, 90, 2: 525–54.

5. Archibald MacLeish, "A Reflection: Riders on Earth Together, Brothers in Eternal Cold," *New York Times*, December 25, 1968, 1.

6. Michael Egan, *Barry Commoner and the Science of Survival: The Remaking of American Environmentalism* (Cambridge, MA: MIT Press, 2007).

7. Jacob Darwin Hamblin, *Poison in the Well: Radioactive Waste in the Oceans at the Dawn of the Nuclear Age* (New Brunswick, NJ: Rutgers University Press, 2008).

8. Linda Lear, *Rachel Carson: Witness for Nature* (New York: Henry Holt, 1997); Priscilla Coit Murphy, *What a Book Can Do: The Publication and Reception of Silent Spring* (Amherst: University of Massachusetts Press, 2005).

9. Published as Kenneth E. Boulding, "The Economics of the Coming Spaceship Earth," in *Environmental Quality in a Growing Economy: Essays from the Sixth Resources for the Future Forum*, edited by Henry Jarrett (Baltimore: The Johns Hopkins University Press, 1966), 3–14.

10. July 1965 speech by Stevenson to United Nations Economic and Social Council, www.adlaitoday.org/articles/connect2_geneva_07-09-65.pdf (accessed November 2011); Richard Buckminster Fuller, *Operating Manual for Spaceship Earth* (Edwardsville: Southern Illinois University Press, 1969); Barbara M. Ward, *Spaceship Earth* (New York: Columbia University Press, 1966). Also, Peder Anker, "Buckminster Fuller as Captain of Spaceship Earth," *Minerva* 2007, 45, 4: 417–34, and R. S. Deese, "The Artifact of Nature: 'Spaceship Earth' and the Dawn of Global Environmentalism," *Endeavour*, 2009, 33, 2: 70–76.

11. Robert Poole, *Earthrise: How Man First Saw the Earth* (New Haven, CT: Yale University Press, 2008); Neil Maher, "Shooting the Moon," *Environmental History* 2004, 9, 3: 526–31. Later reproductions of the picture rotated the image ninety degrees so that it looked as if the earth was actually rising above the moon.

12. MacLeish, "A Reflection."

13. "Fighting to Save the Earth from Man," *Time*, February 2, 1970, 56–63.

14. Matthew Connelly, *Fatal Misconception: The Struggle to Control World Population* (Cambridge, MA: Belknap Press of Harvard University Press, 2008).

15. *Time*, November 8, 1948, 27; William Vogt, *Road to Survival* (London: Victor Gollancz, 1947), 78; also, Thomas Robertson's excellent essay "'This Is the American Earth': American Empire, the Cold War, and American Environmentalism," *Diplomatic History* 2008, 32, 4: 561–84.

16. James Reston, "Feed 'em or Fight 'em," *New York Times*, February 11, 1966, 32.

17. See Garrett Hardin, "The Tragedy of the Commons," *Science* 1968, 162, 3859: 1243–48, and "Living on a Lifeboat," *BioScience* 1974, 24, 10: 561–68.

18. From Barry Commoner, *The Closing Circle: Nature, Man, and Technology* (New York: Knopf, 1971), 297.

19. Chapter 1 of *The Population Bomb* (New York: Ballantine, 1968).

20. Norman Borlaug, December 11, 1970, Nobel Lecture, http://nobelprize.org/nobel_prizes/peace/laureates/1970/borlaug-lecture.html (accessed June 2011).

21. David Anderson, "Mr. Forrester's Terrible Computer," *Wall Street Journal*, September 28, 1971, 18. Background on Forrester and his research comes from several sources including Brian P. Bloomfield, *Modeling the World: The Social Construction of Systems Analysts* (New York: Basil Blackwell, 1986); Thomas Hughes, *Rescuing Prometheus* (New York: Vintage Books, 1998); Fernando Elichirigoity, *Planet Management: Limits to Growth, Computer Simulation, and the Emergence of Global Spaces* (Evanston, IL: Northwestern University Press,

1999); as well as Forrester's own recollections in "System Dynamics—a Personal View of the First Fifty Years," *Systems Dynamics Reviews* 23, 2 and 3: 345–58.

22. Bloomfield's *Modeling the World*, 9–11.

23. Donald Worster, *Nature's Economy: A History of Ecological Ideas* (New York: Cambridge University Press, 1994), notes the parallels between economic and ecological thinking. Also, Joel B. Hagen, *An Entangled Bank: The Origins of Ecosystem Ecology* (New Brunswick, NJ: Rutgers University Press, 1992), and Frank B. Golley, *A History of the Ecosystem Concept in Ecology: More than the Sum of the Parts* (New Haven, CT: Yale University Press, 1993).

24. Paul N. Edwards, *The Closed World: Computers and the Politics of Discourse in Cold War America* (Cambridge, MA: MIT Press, 1996).

25. From the club's original 1970 proposal, "The Predicament of Mankind," submitted to the Volkswagen Foundation; Elichirigoity, *Planet Management*, 60.

26. "Activist on the World Stage: Carroll Wilson Remembered," *Technology Review*, February/March 1984, entrepreneurship.mit.edu/sites/default/files/files/CLW-Activist.pdf (accessed November 2011).

27. Alexander King, "The Club of Rome and Its Policy Impact," in *Knowledge and Power in a Global Society*, edited by William M. Evan (London: SAGE Publications, 1981), 205–24.

28. Forrester quote from Elichirigoity, *Planet Management*, 81.

29. Forrester's methodology is discussed in Paul N. Edwards, "The World in a Machine," in *Systems, Experts, and Computers: The Systems Approach in Management and Engineering, World War II and After*, edited by Agatha C. Hughes and Thomas P. Hughes (Cambridge, MA: MIT Press, 2000), and in Bloomfield's *Modeling the World*.

30. Eduard Pestel, *Beyond the Limits to Growth* (New York: Universe, 1989), 89.

31. "The Dynamics of Global Equilibrium," March 2, 1972, summary adapted from papers by Dennis Meadows, folder 7, box 23, LTG/SIA.

32. *Limits*, 21.

33. From Meadows's 1970 proposal, "Project on the Predicament of Mankind," quoted in Elichirigoity's *Planet Management*, 96.

34. Robert Gillette, "The Limits to Growth: Hard Sell for a Computer View of Doomsday," *Science* 1972, 175, 4026: 1088–92; Peter Passell, Marc Roberts, and Leonard Ross, "Review: The Limits to Growth," *New York Times*, April 2, 1972, BR1.

35. Donella H. Meadows, "The History and Conclusions of *The Limits to Growth*," *Systems Dynamics Review* 2007, 23, 2 and 3: 191–97.

36. Claire Sterling, "A Computer Curve to Doomsday," *Washington Post*, January 5, 1972, A16.

37. "Blueprint for Survival" occupied the entire January 1972 issue of the *Ecologist*, and the principal authors were Edward Goldsmith and Robert Allen; this was republished as a book later that year by Penguin Books.

38. Hal Lindsey with C. C. Carlson, *The Late, Great Planet Earth* (Grand Rapids, MI: Zondervan, 1970). Quotes from Hal Lindsey, *There's a New World*

Coming (Irving, CA: Harvest House, 1973), 93. Also, Paul Boyer, *When Time Shall Be No More: Prophecy Belief in Modern American Culture* (Cambridge, MA: Harvard University Press, 1992).

39. Howard B. Fiske, "Second Coming: There Are Those Who Think It Is Imminent," *New York Times*, October 8, 1972, E8.

40. Description of the conference and general background is in Gillette, "The Limits to Growth."

41. January 24, 1972, and February 11, 1972, letters from Benjamin Read to S. Dillon Ripley and to Anatoliy Dobryin, both in folder 7, box 23, LTG/SIA.

42. Gillette, "The Limits to Growth."

43. *Limits*, 129. Also, "On Reaching a State of Global Equilibrium," *New York Times*, March 13, 1972, 35; "The Dynamics of Global Equilibrium," March 2, 1972, summary adapted from papers by Dennis Meadows, folder 7, box 23, LTG/SIA.

44. William Ophuls, "The Scarcity Society," *Harper's*, April 1974, 47–52.

45. Hardin, "Tragedy of the Commons," 1244.

46. The response to *Limits* produced a huge amount of literature. One perspective is Francis Sandbach, "The Rise and Fall of the *Limits to Growth* Debate," *Social Studies of Science*, 1978, 8, 4: 495–520, while Robert M. Collins, *More: The Politics of Economic Growth in Postwar America* (New York: Oxford University Press, 2000), frames the growth debate in its larger economic context.

47. Anthony Lewis, "To Grow and Die," *New York Times*, January 29, 1972, 29.

48. This was, of course, the subtitle to Schumacher's hugely popular 1973 book *Small is Beautiful*.

49. Leonard Silk, "Predicament of Mankind," *New York Times*, April 12, 1972, 59.

50. Passell, Roberts, and Ross. "Review: The Limits to Growth."

51. David C. Anderson, "A Careful Look at Growth as Suicide," *Wall Street Journal*, March 17, 1972, 6.

52. Economists' reactions are described in chapter 5 of Collins, *More*.

53. Chauncey Starr and Richard Rudman, "Parameters of Technological Growth," *Science*, 1973, 182, 4110: 358–64.

54. One of the earliest and most thorough was H.S.D. Cole et al., eds., *Models of Doom: A Critique of the Limits to Growth* (New York: Universe Books, 1973).

55. Critiques presented in Gillette, "The Limits to Growth."

56. Robert Reinhold, "Mankind Warned of Perils on Growth," *New York Times*, February 27, 1972, 1; Philip H. Abelson, "Limits to Growth," *Science*, 1972, 175, 4027: 1197.

57. Maurice Goldsmith, "Meadows Unlimited or Caveat Computer," *Bulletin of the Atomic Scientists*, 1973, 29, 5: 16–18; Carl Kaysen, "The Computer that Printed Out W*O*L*F*," *Foreign Affairs*, July 1972, 660–68.

58. Elliot L. Richardson, Secretary of Health, Education, and Welfare, quoted in Robert Reinhold, "Warning on Growth Perils Is Examined at Symposium," *New York Times*, March 3, 1972, 41.

59. Ophuls, "The Scarcity Society."

60. George J. Church, "Can the World Survive Economic Growth?" *Time*, August 14, 1972, 56–57.

61. Susan Greenhalgh, "Missile Science, Population Science: The Origins of China's One-Child Policy," *China Quarterly*, 2005, 182, June: 258–59.

62. Collins, *More*, 144.

63. For example, Mihajlo Mesarovic and Eduard Pestel, *Mankind at the Turning Point: The Second Report of the Club of Rome* (New York: E. P. Dutton, 1974).

64. Collins, *More*, 144.

65. Walter Sullivan, "Struggling against Doomsday," *New York Times*, June 11, 1972, E7; Friedel Ungeheuer, "A Stockholm Notebook," *Time*, June 26, 1972, 40.

66. "Environment: Woodstockholm," *Time*, June 19, 1972, 55.

67. Cover of *Newsweek*, November 19, 1973.

68. Quotes from Robert M. Collins, *Transforming America: Politics and Culture in the Reagan Years* (New York: Columbia University Press, 2007), 21.

69. Kaczynski moved to Montana in 1973; while *Soldier of Fortune* appeared in 1975. James Coates, *Armed and Dangerous: The Rise of the Survivalist Right* (New York: Hill and Wang, 1987).

70. From Susan Sontag's "The Imagination of Disaster," *Commentary*, October 1965. Also, Rome, " 'Give Earth a Chance.' "

71. Andrew Ross, *Strange Weather* (New York: Verso, 1991), 141; Vincent Canby, "Movies Are More Sci-Fi than Ever," *New York Times*, March 17, 1974, 113.

72. From Philip Jenkins, *Decade of Nightmares: The End of the Sixties and the Making of Eighties America* (New York: Oxford University Press, 2006), 72; Lance Morrow, "Epitaph for a Decade," *Time*, January 7, 1980, 38–39.

73. Brian Stableford, "Science Fiction and Ecology," in *A Companion to Science Fiction*, edited by David Seed (New York: Blackwell Publishing, 2005), 127–41.

74. William Graebner, "America's Poseidon Adventure: A Nation in Existential Despair," in *America in the Seventies*, edited by Beth Bailey and David Farber (Lawrence: University Press of Kansas, 2004), 157–80.

75. From the 1976 film's opening on-screen preamble; William F. Nolan and George Clayton Johnson, *Logan's Run* (New York: Dial Press, 1967).

Chapter 2: The Inspiration of Limits

1. Al Reinert, "Gerry's World," *Air & Space*, April/May 1989, 28–33.

2. Gerard K. O'Neill, *The High Frontier: Human Colonies in Space* (New York: William Morrow and Company, 1977), 233. Many different editions of this book have appeared since 1977, some with new text while other versions were published in different languages. Unless noted otherwise, I refer to the original version.

3. From Stewart Brand, ed., *Space Colonies: A CoEvolution Book* (San Francisco: Waller Press, 1977), 150; hereafter *Space Colonies*.

4. O'Neill, quoted in ibid., 22.

5. O'Neill, *The High Frontier*, 234–35.

6. This account is largely drawn from "Autobiographical Notes," GKON.

7. Ibid.

8. Reinert, "Gerry's World."

9. Gerard K. O'Neill, "Storage Ring Synchrotron: Device for High Energy Physics Research," *Physical Review* 1956, 102, 5: 1418–19.

10. O'Neill's physics research is described in Elizabeth Paris's doctoral dissertation, "Ringing in the New Physics: The Politics and Technology of Electron Colliders in the United States, 1956–1972," University of Pittsburgh, 1999.

11. "Autobiographical Notes, "GKON.

12. Gerard K. O'Neill, "Storage Rings," *Science* 1963, 141, 3583: 679–86.

13. "Autobiographical Notes," GKON.

14. Undated (c. 1963) note in O'Neill's faculty file, GKON/PUA.

15. Letter, c. April 1970, from O'Neill to Freeman Dyson, GKON.

16. Reinert, "Gerry's World."

17. A 1958 departmental note in O'Neill's faculty file, GKON/PUA.

18. May 10 and July 26, 1967, correspondence between NASA's Manned Space Center and O'Neill as well as July 20, 1967, letter from O'Neill to Walter Bleakney, GKON. W. David Compton, *Where No Man Has Gone Before: A History of Apollo Lunar Exploration Missions*, SP-4214 (Washington, DC: National Aeronautics and Space Administration, 1989).

19. O'Neill, *The High Frontier*, 23.

20. "Interview: Gerard K. O'Neill," *Omni*, July 1979, 76–79, 113–15.

21. Interview with O'Neill in *Cosmic Search Magazine*, 1979, http://www.bigear.org/vol1no2/oneill.htm (accessed November 2008).

22. N. A. Rynin, *K E. Tsiolkovskii: Life, Writings, and Rockets* (Jerusalem: Israel Program for Scientific Translations, 1971), 3.

23. See Asif A. Siddiqi's excellent *The Red Rockets' Glare: Spaceflight and the Soviet Imagination* (New York: Cambridge University Press, 2010).

24. Asif A. Siddiqi, "Making Spaceflight Modern: A Cultural History of the World's First Space Advocacy Group," in *Societal Impact of Space Flight*, edited by Steven J. Dick and Roger D. Launius (Washington, DC: NASA, 2007), 513–38, quote 517.

25. Michael J. Neufeld, "Weimar Culture and Futuristic Technology: The Rocketry and Spaceflight Fad in Germany, 1923–1933," *Technology and Culture* 1990, 31, 4: 725–52.

26. G. Edward Pendray, ed., *The Papers of Robert H. Goddard*, vol. 3 (New York: McGraw-Hill, 1970), 1612.

27. Tom Alexander. "The Wild Birds Find a Corporate Roost." *Fortune*, August 1964, 130–34, 164, 166, 168.

28. Bruce H. Frisch, "Dandridge Cole: G.E.'s Way-out Man," *Science Digest*, July 1965, 9–15. For an example of Dandridge M. Cole's writings see *Beyond Tomorrow: The Next Fifty Years in Space* (Amherst, WI: Amherst Press, 1965).

29. "Outward Bound," *Time*, January 27, 1961, 46.

30. Ibid.

31. Siddiqi, *Red Rockets' Glare*, 16.

32. My description of *Model I* is assembled from a number of sources including O'Neill's own notes from 1972–73 as well as drafts of his articles and talks, all GKON. A fuller description is in Gerard K. O'Neill, "The Colonization of Space," *Physics Today* 1974, 27, 9: 32–40.

33. O'Neill, *The High Frontier*, 210.

34. April 15, 1972, letter from James C. Fletcher to Sen. Walter Mondale; from John M. Logsdon et al., eds., *Exploring the Unknown: Selected Documents in the History of the U.S. Civil Space Program*, vol. 4, *Accessing Space* (Washington, DC: NASA, 1999).

35. From O'Neill's July 23, 1975, testimony to the congressional Subcommittee on Space Science and Applications; reprinted in Brand, *Space Colonies*.

36. Gerard K. O'Neill, "A Lagrangian Community?" *Nature* 1974, 250, 5468: 636.

37. O'Neill, *The High Frontier*, 242.

38. Henry H. Kolm and Richard D. Thornton, "Electromagnetic Flight," *Scientific American* 1973, 229, 4: 17–25.

39. Arthur C. Clarke, "Electromagnetic Launching as a Major Contribution to Space-Flight," *Journal of the British Interplanetary Society* 1950, 9, 6: 261–67.

40. Anonymous reviews of "The Colonization of Space" manuscript from *Science*, undated but sometime mid-1972, GKON.

41. May 26, 1972, letter from O'Neill to George Pimentel, GKON.

42. George Dyson, *Project Orion: The Atomic Spaceship, 1957–1965* (New York: Allen Lane Science, 2002).

43. Freeman J. Dyson, "Death of a Project," *Science* 1965, 149, 3680: 141–44.

44. Dyson's introduction to Gerard K. O'Neill, *The High Frontier: Human Colonies in Space*, 3rd ed. (Burlington, Ontario: Apogee Books, 2000), 5–7.

45. Dyson, *Project Orion*, 186–87.

46. April 11, 1972, letter from Dyson to O'Neill, GKON.

47. J. D. Bernal, *The World, the Flesh, and the Devil: An Enquiry into the Three Enemies of the Rational Soul*, 2nd ed. (Bloomington: Indiana University Press, 1969). Described in Maurice Goldsmith, *Sage: A Life of J. D. Bernal* (London: Hutchinson, 1980). See, also, essays in Brenda Swann and Francis Aprahamian, eds., *J .D. Bernal: A Life in Science and Politics* (New York: Verso, 1999), and Andrew Brown, *J. D. Bernal: The Sage of Science* (New York: Oxford University Press, 2005). Bernal's book was part of a British series on the future of science and technology called Today and Tomorrow that also included offerings from J.B.S. Haldane and Bertrand Russell.

48. Quotes from Bernal, *The World, the Flesh, and the Devil*.

49. Dyson's lecture was published as Freeman J. Dyson, "The World, the Flesh and the Devil," in *Communication with Extraterrestrial Intelligence*, edited by Carl Sagan (Cambridge, MA: MIT Press, 1972), 370–89.

50. Dyson, "The World, the Flesh and the Devil," 373.

51. Dyson, "The World, the Flesh and the Devil," 375–77, 383.

52. Freeman J. Dyson, "Human Consequences of the Exploration of Space," *Bulletin of the Atomic Scientists* 1969, 25, 7: 8–13.

53. April 11, 1972, letter from Dyson to O'Neill, GKON.

54. April 20, 1972, letter from O'Neill to Dyson, GKON.

55. Kenneth Brower, *The Starship and the Canoe* (New York City: Bantam Books, 1978), 159.

56. O'Neill's recollections, recounted in *The High Frontier.*

57. Handwritten draft of "Newsletter on Colonization of Space," December 24, 1973, GKON.

58. The history of Brand, *Whole Earth*, and the Point Foundation are covered in Andy Kirk, *Counterculture Green: The Whole Earth Catalog and American Environmentalism* (Lawrence: University of Kansas Press, 2007).

59. Phillips, quoted in O'Neill, *The High Frontier*, 248.

60. The 1974 talks are presented as "Proceedings of the Princeton Conference on the Colonization of Space; May 10, 1974," appendix in *Space Manufacturing Facilities: Proceedings of the Princeton/AIAA/NASA Conference*, edited by Jerry Gray (New York: American Institute of Aeronautics and Astronautics, 1977).

61. May 3, 1974, press release, Princeton University News Bureau, GKON/PUA.

62. Walter Sullivan, "Proposal for Human Colonies in Space Is Hailed by Scientists as Feasible Now," *New York Times*, 1974, A1.

63. Gerard K. O'Neill, "The Colonization of Space," *Physics Today* 1974, 27, 9: 32–40. While the *Physics Today* article was the first major peer-reviewed presentation of O'Neill's ideas, other pieces describing space colonies appeared in the summer of 1974 including O'Neill, "A Lagrangian Community?" and Paul T. Libassi, "Space to Grow," *Sciences*, July/August 1974, 15–20.

64. O'Neill, "Colonization of Space," 32, emphasis in original.

65. De Witt Douglas Kilgore, *Astrofuturism: Science, Race, and Visions of Utopia in Space* (Philadelphia: University of Pennsylvania Press, 2003).

66. O'Neill, *The High Frontier*, 23.

67. Here I am drawing on several works including Frank E. Manuel and Fritzie P. Manuel, *Utopian Thought in the Western World* (Cambridge, MA: Harvard University Press, 1979); chapter 5 of John F. Kasson's *Civilizing the Machine: Technology and Republican Values in America, 1776–1900* (New York: Grossman, 1976); and Howard P. Segal's *Technological Utopianism in American Culture* (Chicago: University of Chicago Press, 1985).

68. O'Neill, *The High Frontier*, 200–201; O'Neill, "Colonization of Space, 36.

69. Robert H. Kargon and Arthur P. Molella, *Invented Edens: Techno-Cities of the Twentieth Century* (Cambridge, MA: MIT Press, 2008).

70. Peder Anker, "The Ecological Colonization of Space," *Environmental History* 2005, 10, 2: 239–68.

71. Leo Marx, *The Machine in the Garden: Technology and the Pastoral Ideal in America* (New York: Oxford University Press, 1964); also, see Howard Segal, "Leo Marx's 'Middle Landscape': A Critique, a Revision, and an Appreciation," *Reviews in American History* 1977, 5, 1: 137–50.

72. Daniel Bell, *The Coming of Post-Industrial Society: A Venture in Social Forecasting* (New York: Basic Books, 1973), 487.

73. Published in Murray Bookchin, *Post-Scarcity Anarchism* (Berkeley, CA: Ramparts Press, 1971).

74. Frederick Jackson Turner, "The Significance of the Frontier in American

History," in *Rereading Frederick Jackson Turner*, edited by John M. Farager (New York: Henry Holt, 1994), 31–60.

75. Patricia Nelson Limerick, "Imagined Frontiers: Westward Expansion and the Future of the Space Program," in *Space Policy Alternatives*, edited by Radford Byerly (Boulder, CO: Westview Press, 1992), 249–61; Howard E. McCurdy, *Space and the American Imagination:* (Washington, DC: Smithsonian Institution Press, 1997), chapter 6.

76. Gerard O'Neill, "Space—Our New American Frontier," unpublished essay, undated but likely from the mid-1980s, GKON.

77. O'Neill, quoted in Sullivan, "Proposal for Human Colonies in Space," A1, 23.

78. David E. Nye, *America as Second Creation: Technology and Narratives of New Beginnings* (Cambridge, MA: MIT Press, 2003).

79. Ibid., 1–2.

Chapter 3: Building Castles in the Sky

1. Gerard K. O'Neill, "The Colonization of Space," *Physics Today* 1974, 27, 9: 39.

2. Based on O'Neill's own recollections as well as those of Tasha O'Neill. Unfortunately, neither O'Neill's personal papers nor his official Princeton records preserved this correspondence.

3. "Newsletter in Space Colonization Studies," January 9, 1975, SSI.

4. Isaac Asimov, "First Colony in Space," *National Geographic*, July 1976, 76–89.

5. Chris Dubbs and Emeline Paat-Dahlstrom, *Realizing Tomorrow: The Path to Private Spaceflight* (Lincoln: University of Nebraska Press, 2011), 24.

6. "Lagrangia: Pioneering in Space," *Science News* 1974, 106, 12: 183. The term appears in other places including a 1974 profile of O'Neill; Graham Chedd, "Colonisation at Lagrangea," *New Scientist*, October 24, 1974, 247–49.

7. Donald Malcolm, "Notes and Comments," *New Yorker*, June 17, 1974, 23. Chapter 5 of Kilgore's *Astrofuturism* offers a critique of the parallels between suburbia and O'Neill's hypothetical settlements.

8. Chedd, "Colonisation at Lagrangea," 249.

9. Ed Regis, *Nano! Remaking the World Atom by Atom* (Boston: Little, Brown, 1995), 89.

10. Jonathan Allan, ed., *March 4: Scientists, Students, and Society* (Cambridge, MA: MIT Press, 1970).

11. *Massachusetts Institute of Technology Bulletin*, September 1973, 108, 3: 191–92

12. Barbara Marx Hubbard, *The Hunger of Eve: One Woman's Odyssey toward the Future*, 2nd ed. (Eastbound, WA: Sweet Forever Publishing, 1989 [originally published in 1976]), 106.

13. Hubbard, *Hunger of Eve*, 93.

14. Described in Hubbard, *Hunger of Eve*, chapter 5.

15. William Sims Bainbridge, *The Spaceflight Revolution: A Sociological Study* (Malabar, FL: Robert E. Krieger Publishing, 1983); Roger D. Launius, "Perfect Worlds, Perfect Societies: The Persistent Goal of Utopia in Human Spaceflight," *Journal of the British Interplanetary Society* 2003, 56: 338–49.

16. Bruce J. Schulman, *The Seventies: The Great Shift in American Culture, Society, and Politics* (New York: Da Capo Press, 2001), 92–101; Lynn Smith, "Futurist Spreads Her Message of Unlimited Human Potential," *Los Angeles Times*, February 6, 1985, D1.

17. Krafft Ehricke, "The Extraterrestrial Imperative," *Bulletin of the Atomic Scientists* 1971, 27, 9: 18–26.

18. MIT presented the class's final product as "A Systems Design for a Prototype Space Colony," dated Spring 1976; copy in author's files.

19. For example, see the essays collected in Charles H. Holbrow, Allan M. Russell, and Gordon Sutton, F., eds., *Space Colonization: Technology and the Liberal Arts* (New York: American Institute of Physics, 1986).

20. Mark S. Miller, July 24, 2009, interview with the author.

21. Mark M. Hopkins, August 24, 2009, interview with the author; also Michael A. G. Michaud, *Reaching for the High Frontier: The American Pro-Space Movement, 1972–1984* (New York: Praeger, 1986), 82–83.

22. From the event's official brochure and list of press attendees, GKON/PUA; also, Walter Sullivan, "Princeton Gathering Makes Detailed Assessment of Problems in Establishing a Colony of 10,000 in Space," *New York Times*, May 12, 1975, 42.

23. Mark M. Hopkins, August 24, 2009, interview with the author.

24. From the study's formal announcement, included in O'Neill's "Newsletter in Space Colonization Studies," January 9, 1975, SSI.

25. Mark M. Hopkins, August 24, 2009, interview with the author.

26. November 10, 1975, letter to "Harry" (probably Henry Kolm) from O'Neill, GKON.

27. Dubbs and Paat-Dahlstrom, *Realizing Tomorrow*, 15–18.

28. "Space Sciences," *Science News*, November 13, 1976, 314.

29. Sandra Blakeslee, "Experts Press for Space Colony Now," *New York Times*, August 23, 1975, 47.

30. NASA Press Release 75-249 and "Summary of Report of the 1975 NASA/Stanford Summer Study of Space Colonization," GKON/PUA. Also, "Space Colonies: Home, Home on Lagrange," *Science News* 1975, 109, 10: 149. NASA eventually published the 1975 study as a technical report: Richard D. Johnson and Charles Holbrow, eds., *Space Settlements: A Design Study*, SP-413 (Washington, DC: National Aeronautics and Space Administration, 1977).

31. *Trenton Times*, August 4, 1975, clipping in GKON/PUA.

32. From *Space Resources and Space Settlements*, SP-428 (Washington, DC: National Aeronautics and Space Administration, 1979).

33. O'Neill, *The High Frontier*, 256.

34. Peter E. Glaser, "Power from the Sun: Its Future," *Science* 1968, 162, 3856: 857–61.

35. William Brown, "The Amplitron: A Super Power Microwave Generator," *Electronics Progress* 1960, 5, 1: 1–5.

36. Peter E. Glaser., "Method and Apparatus for Converting Solar Radiation to Electrical Power," patent 3,781,647, issued by United States Patent and Trademark Office, December 25, 1973.

37. Gerard K. O'Neill, "Space Colonies and Energy Supply to the Earth," *Science* 1975, 190, 4218: 943–47.

38. Anthony Lewis, "Nearing the Limits: II," *New York Times*, October 4, 1973, 45.

39. "Transcript of President's Address on Energy Situation," *New York Times*, November 8, 1973, 32.

40. "Transcript of State of the Union Address," *New York Times*, January 31, 1974, 20.

41. "Science: Colonizing Space," *Time*, May 26, 1975, 60.

42. Outlook for Space Study Group, *Outlook for Space: Report to the NASA Administrator*, SP-386 (Washington, DC: National Aeronautics and Space Administration, 1976).

43. Subcommittee on Space Science and Applications of the Committee on Science and Technology, *Future Space Programs 1975*, first session, 94th Congress, July 22–24, 29–30, 1975, p. 3.

44. Dubbs and Paat-Dahlstrom, *Realizing Tomorrow*, 23.

45. Ed Regis, *Great Mambo Chicken and the Transhuman Condition: Science Slightly over the Edge* (New York: Penguin Books, 1990), 57. Additional biographical information on the Hensons comes from a number of sources including chapter 5 of Michaud, *Reaching for the High Frontier*, and several in-person and phone interviews including H. Keith Henson (May 2, 2007), James C. Bennett (July 19, 2007), Conrad Schneiker (November 9, 2007), and Mark M. Hopkins, (August 24, 2009); interview notes in author's collection.

46. Robin Snelson, "Space Now! An Interview with Keith and Carolyn Henson," *Future*, August 1978, 53–57.

47. Tom Wolfe, "The 'Me' Decade and the Third Great Awakening," *New York*, August 23, 1976, 26–40; Wade C. Roof, *A Generation of Seekers: The Spiritual Journeys of the Baby Boom Generation* (New York: HarperCollins, 1993).

48. Snelson. "Space Now!" 57.

49. *L5 News*, September 1975, 2.

50. Trudy Bell, "American Space-Interest Groups," *Star & Sky*, September 1980, 53–60, and "Space Activism," *Omni*, February 1981, 50–54. Also, Trudy Bell, August 8, 2008, phone interview with the author.

51. Sam Binkley, *Getting Loose: Lifestyle Consumption in the 1970s* (Durham, NC: Duke University Press, 2007).

52. *L5 News*, November 1977, 9.

53. Taylor Dark III, March 11, 2010, interview with the author.

54. James C. Bennett, July 19, 2007, interview with the author. A version of this space-libertarian ideology can be found in Edward L. Hudgins, ed., *Space: The Free-Market Frontier* (Washington, DC: Cato Institute, 2002).

55. James C. Bennett, July 19, 2007, interview with the author.

56. Hugh Millward, "Where Is the Interest in Space Settlement?" *L5 News*, January 1980, 8–10.

57. Michaud, *Reaching for the High Frontier*, 112–14.

58. Ruth Oldenziel, *Making Technology Masculine: Men, Women, and Modern Machines in America, 1870–1945* (Amsterdam: Amsterdam University Press, 1999).

59. *L5 News*, February 1977, 18.

60. "Home on Lagrange (The L5 Song)," copyright 1978 by William S. Higgins and Barry D. Gehm; the lyrics originally appeared in *CoEvolution Quarterly*'s Summer 1978 issue; reprinted in July 1979 issue of *L5 News*.

61. Personal correspondence between the author and Higgins and Gehm, April 6, 2012.

62. Mark M. Hopkins, August 24, 2009, interview with the author; Nicholas Wade, "Limits to Growth: Texas Conference Finds None, but Didn't Look Too Hard," *Science* 1975, 190, 4214: 540–41.

63. J. Peter Vajk, *Doomsday Has Been Cancelled* (Culver City, CA: Peace Press, 1978), xiv–xv.

64. Aden Baker Meinel and Majorie Pettit Meinel, "Physics Looks at Solar Energy," *Physics Today*, February 1972, 44–50; Frank N. Laird, "Constructing the Future: Advocating Energy Technologies in the Cold War," *Technology and Culture* 2003, 44, 1: 27–49.

65. *L5 News*, October 1975, 4.

66. *L5 News*, July 1976, 1–3.

67. Fred Turner, *From Counterculture to Cyberculture: Stewart Brand, the Whole Earth Network, and the Rise of Digital Utopianism* (Chicago: University of Chicago Press, 2006).

68. Michaud, *Reaching for the High Frontier*, 88.

69. October 17, 1980, letter from O'Neill to John F. McCarthy, director of NASA's Lewis Research Center in Ohio, GKON; I am grateful for publications and reports on the MIT-Princeton mass-driver work that Henry Kolm shared with me.

70. A long discussion of O'Neill's ideas appeared in a two-part episode of *Nova*, produced by public television station WGBH in Boston, that first aired in January 1978.

71. Recollections from discussions and e-mail exchanges with Tasha O'Neill and Freeman Dyson.

72. "Newsletter on Space Studies at Princeton," November 1976, 2, GKON.

73. Carolyn Henson, "A Message from the Ex-President," *L5 News*, December 1979, 3.

74. Timothy Leary, *Starseed* (San Francisco: Level Press, 1973), http://www.lycaeum.org/books/books/starseed/starseed.shtml (accessed November 2010). A version of this essay was reprinted in Leary's 1977 book *Neuropolitics*.

75. Ibid.

76. Flyer for April 16–17, 1977, gathering by Wilson, KH-AL/SA. Also described in Wilson's book *Prometheus Rising* (Las Vegas: New Falcon Publications, 1983).

77. Described at http://www.cryonics.org/luna.html (accessed May 2011).

78. Timothy Leary, *Flashbacks: An Autobiography* (Los Angeles: J. P. Tarcher, 1983), 365.

79. William Overend, "Timothy Leary: Messenger of Evolution," *Los Angeles*

Times, January 30, 1977; Robin Snelson, "Interview: Timothy Leary," *Future Life*, May 1979, 33–34, 66.

80. Elizabeth Robinson, "Movement into Space: A View from Two Worlds, Pt. 2," *L5 News*, January 1977, 3.

81. Described in the October 1976 issue of *L5 News*.

82. "George Koopman Dies in Wreck, Technologist for Space Was 44," *New York Times*, July 21, 1989, B8.

83. David Kaiser, *How the Hippies Saved Physics: Science, Counterculture, and the Quantum Revival* (New York: W. W. Norton & Co., 2011).

84. Douglas Martin. "George Leonard, Voice of '60s Counterculture Dies at 86." *New York Times*, January 18, 2010, A22.

85. Timothy Leary with contributions from Robert Anton Wilson and George Koopman, *Neuropolitics: The Sociobiology of Human Metamorphosis* (Los Angeles: Starseed/Peace Press, 1977).

86. Ibid., 49; this essay first appeared in the August 1976 issue of *L5 News*.

87. Ibid., 100.

88. Ibid., 83.

89. Overend, "Timothy Leary," E1.

90. *L5 News*, September 1976, 19.

91. Timothy Leary, "Scientist Superstar.," *Future Life*, February 1981, 70–73.

92. Stewart Brand, ed., *Space Colonies: A CoEvolution Book* (San Francisco: Waller Press, 1977), 8.

93. January 21, 1976, diary note from 1975–76 notebooks, folder 6, box 18, SB/SA.

94. Francis French and Colin Burgess, *In the Shadow of the Moon: A Challenging Journey to Tranquility, 1965–1969* (Lincoln: University of Nebraska Press, 2007), 334. Also, Russell L. Schweickart, July 24, 2008, interview with the author.

95. Brand, *Space Colonies*, 6.

96. From Andy Kirk's *Counterculture Green: The Whole Earth Catalog and American Environmentalism* (Lawrence: University of Kansas Press, 2007), 164; chapter 5 details Brand's *CoEvolution Quarterly* years including the space colonies debate. Also, Peder Anker, "The Ecological Colonization of Space," *Environmental History* 2005, 10, 2: 239–68.

97. September 15, 1976, transcript of Brand's talk at World Game Workshop, folder 6, box 7, WEC/SA.

98. January 25, 1975, letter from Brand to Morrison, folder 5, box 2; and June 24, 1975, draft letter from Brand, folder 9, box 5; both WEC/SA.

99. Undated 1978 letter. folder 9, box 12, WEC/SA.

100. Carroll Pursell, "The Rise and Fall of the Appropriate Technology Movement in the United States, 1965–1985," *Technology and Culture* 1993, 34, 3, 629–27.

101. Kirk, *Counterculture Green*, 54–55.

102. Michael Allen, " 'I Just Want to Be a Cosmic Cowboy': Hippies, Cowboy Code, and the Culture of a Counterculture," *Western Historical Quarterly* 2005,

36, Autumn, 275–99; quotes from Stewart Brand, "Free Space," *CoEvolution Quarterly*, Fall 1975, 4.

103. Brand, "Free Space," 4.

104. Brand, *Space Colonies*, 3.

105. Stewart Brand, mid-1977 journal entry, SB/SA.

106. Brand, *Space Colonies*, 33, 42.

107. *CoEvolution Quarterly*, Spring 1976, 5, 47, 48, 52.

108. *CoEvolution Quarterly*, Spring 1976, 23.

109. November 29, 1976, letter from Holt to Brand, folder 15, box 6, WEC/SA.

110. June 4, 1976, letter from Berry, cc'ed to Brand, folder 3, box 4, WEC/SA.

111. *CoEvolution Quarterly*, Spring 1976, 8.

112. June 18, 1976, letter from Gurney Norman to Brand, folder 3, box 7, WEC/SA.

113. December 1, 1977, letter from Diane Engle to Brand, folder 11, box 1, WEC/SA.

114. *CoEvolution Quarterly*, Spring 1976, 5.

115. Jib Fowles, "Review—*The High Frontier: Human Colonies in Space*," *Technology and Culture* 1977, 18, 4: 718–20.

116. Jay S. Huebner, "Teaching about the Colonization of Space," *American Journal of Physics* 1979, 47, 3: 228–31; Kaiser, *How the Hippies Saved Physics*.

117. *SSI Update*, May/June 1992, 2, SSI.

118. According to SSI records, materials in author's file.

119. "Out of This World," segment broadcast by *60 Minutes* on October 16, 1977, produced by Imre Horvath and narrated by Dan Rather.

120. *L5 News*, December 1977, 17; "Presidential Directive/NSC 42: Civil and Further National Space Policy," October 10, 1978, memo; in John M. Logsdon et al., eds., *Exploring the Unknown: Selected Documents in the History of the U.S. Civil Space Program*, vol. 1, *Organizing for Exploration* (Washington, DC: National Aeronautics and Space Administration, 1995).

121. George Alexander, "Proposed State Use of Satellite Detailed," *Los Angeles Times*, December 23, 1977, A3.

122. Roger Rapoport, *California Dreaming: The Political Odyssey of Pat & Jerry Brown* (Berkeley, CA: Nolo Press, 1982), 204.

123. From November 21 and 28, 1977, letters to Brown and Brand, folder 3, box 12, WEC/SA.

Chapter 4: Omnificent

1. Robert D. McFadden, "Bob Guccione, 79, Dies," *New York Times*, October 22, 2010, A34.

2. Patricia Bosworth, "The X-Rated Emperor," *Vanity Fair*, February 2005, 148–61; Robin Pogrebin, "Kathy Keeton Guccione, 58, President of Magazine Company," *New York Times*, September 23, 1997, D27.

3. Philip H. Dougherty, "Advertising: Guccione's New Aim with Omni," *New York Times*, August 2, 1979, D13.

4. Trudy E. Bell, August 8, 2008, phone interview with the author.

5. Arlie Schardy, "The Science Boom," *Newsweek*, September 17, 1979, 104.

6. Frederic Golden, "The Cosmic Explainer," *Time*, October 20, 1980, 62–69.

7. Bruce V. Lewenstein, "Was There Really a Popular Science 'Boom'?" *Science, Technology, and Human Values* 1987, 12, 2: 29–41.

8. Schardy, "The Science Boom."

9. Allen Hammond, editor of *Science 80*, quoted in Lewenstein, "Was There Really a Popular Science 'Boom'?"

10. Bob Guccione, *Omni*, October 1978, 6.

11. Fred Turner, "Where the Counterculture Met the New Economy: Revisiting the WELL and the Origins of Virtual Community," *Technology and Culture* 2005, 46, 3: 485–512.

12. K. Eric Drexler, "Exploring Future Technologies," in *The Reality Club*, edited by John Brockman (New York: Lynx Books, 1988), 132–33.

13. Gene Bylinsky and Zhenya Lane, "Wizards of Silicon Valley," *Omni*, August 1979, 54–58, 119–20.

14. Christophe Lécuyer, *Making Silicon Valley: Innovation and the Growth of High Tech, 1930–1970* (Cambridge, MA: MIT Press, 2006); Ross Knox Bassett, *To the Digital Age: Research Labs, Start-up Companies, and the Rise of MOS Technology* (Baltimore: The Johns Hopkins University Press, 2002).

15. Bylinsky and Lane, "Wizards of Silicon Valley."

16. Michael Riordan and Lillian Hoddeson, *Crystal Fire: The Invention of the Transistor and the Birth of the Information Age* (New York: W. W. Norton & Co., 1997).

17. Richard P. Feynman, "How to Build an Automobile Smaller than This Dot," *Popular Science*, November 1960, 114–16, 230–32.

18. Richard P. Feynman, "There's Plenty of Room at the Bottom," *Engineering and Science*, February 1960, 22–36; also, chapter 4 of Ed Regis, *Nano! Remaking the World Atom by Atom* (Boston: Little, Brown, 1995).

19. For example, "The Feynman Awards," *Time*, December 12, 1960:,50, 52. Chris Toumey, "Apostolic Succession," *Engineering & Science* 2005, 1/2: 16–23; Andreas Junk and Falk Riess, "From an Idea to a Vision: There's Plenty of Room at the Bottom," *American Journal of Physics* 2006, 74, 9: 825–30.

20. "New Miniaturization Award," *Science* 1957, 126, 3279: 922.

21. Horace D. Gilbert, ed., *Miniaturization* (New York: Reinhold Publishing Corporation, 1961), 12.

22. J. A. Morton and W. J. Pietenpol, "The Technological Impact of Transistors," *Proceedings of the IRE* 1958, 46, 6: 955–59.

23. Thomas J. Misa, "Military Needs, Commercial Realities, and the Development of the Transistor, 1948–1958," in *Military Enterprise and Technological Change: Perspectives on the American Experience*, edited by Merritt Roe Smith (Cambridge, MA: MIT Press, 1985), 253–87.

24. Christophe Lécuyer and David C. Brock, *Makers of the Microchip: A Documentary History of Fairchild Semiconductor* (Cambridge, MA: MIT Press, 2010).

25. David C. Brock, ed., *Understanding Moore's Law: Four Decades of Innovation* (Philadelphia: Chemical Heritage Foundation, 2006), 31.

26. From the biographical sketch that accompanied Moore's "Cramming More Components onto Integrated Circuits," *Electronics* 1965, 38, 8: 114–17.

27. Moore later revised his doubling timescale to every few years, but the validity of his initial observation remained.

28. Paul E. Ceruzzi, "Moore's Law and Technological Determinism: Reflections on the History of Technology," *Technology and Culture* 2005, 46, 3: 584–93; Ethan Mollick, "Establishing Moore's Law," *IEEE Annals of the History of Computing* 2006, 28, 3: 62–75.

29. Victor K. McElheny, "Revolution in Silicon Valley," *New York Times*, June 20, 1976, 97.

30. Ibid.; Pamela G. Hollie, "To Intel Founder, Medal Is a Sign," *New York Times*, February 18, 1980, D3.

31. A rich portrait of excitement and concerns raised circa 1980 can be found in Tom Forester, ed., *The Microelectronics Revolution: The Complete Guide to the New Technology and Its Impact on Society* (Cambridge, MA: MIT Press, 1981).

32. D. C. Englebart, "Augmenting Human Intellect: A Conceptual Framework," Report to the Director of Information Sciences, Air Force Office of Scientific Research, Stanford Research Center, October 1962,http://www.dougen gelbart.org/pubs/augment-3906.html (accessed June 2011).

33. Engelbart, quoted in *Interactive Media*, S. Ambron and K. Hooper, eds. (Redmond: Microsoft Press, 1988), 19. Also, Thierry Bardini, *Bootstrapping: Douglas Engelbart, Coevolution, and the Origins of Personal Computing* (Stanford, CA: Stanford University Press, 2000).

34. Engelbart's demo is available at http://sloan.stanford.edu/MouseSite/1968 Demo.html (accessed June 2011).

35. Alan C. Kay, from 1989 talk titled "Predicting the Future," *Stanford Engineering*, Autumn 1989, 1–6.

36. Stewart Brand, "Spacewar: Fanatic Life and Symbolic Death among the Computer Bums," *Rolling Stone*, December 7, 1972, 50–58.

37. From the January 1975 cover of *Popular Electronics*. The history of the personal computer is extensive; one succinct introduction is Paul E. Ceruzzi, "From Scientific Instrument to Everyday Appliance: The Emergence of Personal Computers, 1970–77," *History and Technology* 1996, 13, 1: 1–31.

38. Ted Nelson, "Computer Lib," *Omni*, November 1978, 57–62.

39. See chapter 21 of Noah Wardrip-Fruin and Nick Montfort, eds., *The New Media Reader* (Cambridge, MA: MIT Press, 2003).

40. Theodor Holm Nelson, *Computer Lib/Dream Machines* (self-published, 1974).

41. Ted Nelson, "Computer Lib," *Omni*, November 1978, 57–62.

42. For example, Robert N. Noyce, "Microelectronics," *Scientific American*, March 1977, 63–69.

43. Paul E. Ceruzzi, "Moore's Law and Technological Determinism: Reflections of the History of Technology," *Technology and Culture* 2005, 46, 3: 584–93.

44. Kathleen and Sharon McAuliffe, "The Gene Trust," *Omni*, March 1980, 62–66, 120–22.

45. For background, Horace Freeland Judson, *The Eighth Day of Creation: Makers of the Revolution in Biology* (New York: Touchstone Books, 1979); Martin Kenney, *Biotechnology: The University Industrial Complex* (New Haven, CT: Yale University Press, 1988); Sheldon Krimsky, *Biotechnics and Society: The Rise of Industrial Genetics* (Westport, CT: Praeger Publishing, 1991); Michel Morange and Matthew Cobb, *A History of Molecular Biology* (Cambridge, MA: Harvard University Press, 1998).

46. For example, Victor K. McElheny, "Gene Transplants Seen Helping Farmers and Doctors," *New York Times*, May 20, 1974, 61; Gene Bylinski, "Industry Is Finding More Jobs for Microbes," *Fortune*, February 1974, 96–102.

47. Sheldon Krimsky, *Genetic Alchemy: The Social History of the Recombinant DNA Controversy* (Cambridge, MA: MIT Press, 1982); Susan Wright, *Molecular Politics: Developing American and British Regulatory Policy for Genetic Engineering, 1972–1982* (Chicago: University of Chicago Press, 1994).

48. Dennis Bray, *Wetware: A Computer in Every Living Cell* (New Haven, CT: Yale University Press, 2010).

49. From the December 10, 1965, award ceremony speech in Stockholm, http://nobelprize.org/nobel_prizes/medicine/laureates/1965/press.html. Monod and Jacob shared the prize with microbiologist André Lwoff.

50. Lily E. Kay, *Who Wrote the Book of Life? A History of the Genetic Code* (Stanford, CA: Stanford University Press, 2000).

51. From the Web exhibit on Nirenberg presented by the National Institutes of Health, http://profiles.nlm.nih.gov/ (accessed January 2011).

52. Terms for the technique varied; see James F. Danielli, "Artificial Synthesis of New Life Forms," *Bulletin of the Atomic Scientists* December 1972, 20–24, while Robert L. Sinsheimer, the head of Caltech's Division of Biology, used the term "synthetic biology" in his "Recombinant DNA—on Our Own," *BioScience* 1976, 26, 10: 599.

53. McElheny, "Gene Transplants."

54. Chemist Arne Tiselius quoted in William L. Laurence. "Structure of Life." *New York Times*, January 14, 1962, E7.

55. Krimsky, *Genetic Alchemy*, 294–311.

56. Frederic Golden. "Shaping Life in the Lab." *Time*, March 8, 1981: 50–56.

57. McElheny, "Gene Transplants."

58. Sally Smith Hughes, "Making Dollars out of DNA: The First Major Patent in Biotechnology and the Commercialization of Molecular Biology, 1974–1980," *Isis* 2001, 92, 4: 541–75, as well as her book *Genentech: The Beginnings of Biotech* (Chicago: University of Chicago Press, 2011).

59. Daniel J. Kevles, "Diamond vs. Chakrabarty and Beyond: The Political Economy of Patenting Life," in *Private Science: Biotechnology and the Rise of the Molecular Sciences*, edited by Arnold Thackray (Philadelphia: University of Pennsylvania Press, 1998), 65–79.

60. Proponents of this alleged neologism ignored the fact that European scientists labeled research associated with brewing and fermentation as *biotechnologie* decades earlier; Robert Bud, "Molecular Biology and the Long-Term History of Biotechnology," in *Private Science: Biotechnology and the Rise of the Molecular*

Sciences, edited by Arnold Thackray (Philadelphia: University of Pennsylvania Press, 1998), 3–19. "Frenetic Engineering," *Economist*, October 18, 1980, 108.

61. Robert Teitelman, *Gene Dreams: Wall Street, Academia, and the Rise of Biotechnology* (New York: Basic Books, 1989), 27 and 35. Also, chapters 8 and 9 of Robert Bud, *The Uses of Life: A History of Biotechnology* (New York: Cambridge University Press, 1993), and Paul Rabinow, *Making PCR: A Story of Biotechnology* (Chicago: University of Chicago Press, 1996).

62. Nicholas Wade, "Gene Splicing Company Wows Wall Street," *Science* 1980, 210, 4469: 506–7; Thomas Lueck, "Cetus in Record Offering; Market Response Is Cool," *New York Times*, March 7, 1981, 31. Also, Hughes, *Genentech*.

63. Kathleen and Sharon McAuliffe, "The Gene Trust," *Omni*, March 1980, 122.

64. *Omni*, December 1978, 10.

65. Trudy E. Bell, August 8, 2008, phone interview with the author.

66. Trudy E. Bell, July 15, 2011, personal communication with the author.

67. Ben Bova, June 29, 2007, phone interview with the author.

68. Sam Binkley, "The Seers of Menlo Park: The Discourse of Heroic Consumption in the 'Whole Earth Catalog,'" *Journal of Consumer Culture* 2003, 3, 3: 283–313.

69. *Omni*, October 1979, 40–41.

70. For an interesting parallel, see Kenon Breazeale, "In Spite of Women: *Esquire* Magazine and the Construction of the Male Consumer," *Signs* 1994, 20, 1: 1–22.

71. *Omni*, advertising copy, March 1979 and November 1979, 46.

72. Before her death in 1997, Keeton also published a book called *Longevity: The Science of Staying Young*.

73. David Kaiser, *How the Hippies Saved Physics: Science, Counterculture, and the Quantum Revival* (New York: W. W. Norton & Co., 2011).

74. Ben Bova, June 29, 2007, phone interview with the author.

75. Ibid.

76. Orson Scott Card, "A Thousand Deaths," *Omni*, December 1978, 98–100, 136–40.

77. For example, "Interview: Gerard K. O'Neill," *Omni*, July 1979, 77–79, 113–15.

78. *Omni*, June 1983, 6; *Omni*, December 1980, 6.

79. McFadden, "Bob Guccione, 79, Dies."

80. The full name of the treaty was the "Agreement Governing the Activities of States on the Moon and Other Celestial Bodies"; the quote comes from *Omni*, November 1979, 6.

81. Patricia Seremet, "Last Word," *Omni*, August 1980, 130.

82. *Omni*, January 1979, 10.

83. Richard Barbrook and Andy Cameron, "The California Ideology," *Science as Culture* 1996, 6, 6: 44–72.

84. Chapter 7 of Fred Turner's *From Counterculture to Cyberculture: Stewart Brand, the Whole Earth Network, and the Rise of Digital Utopianism* (Chicago: University of Chicago Press, 2006) presents an excellent view on the *Wired* era.

85. *Omni*, December 1981, 14.

86. Newt Gingrich, *Window of Opportunity: A Blueprint for the Future* (New York: Tor Books, 1984); John B. Judis, "Newt's Not-So-Weird Gurus," *New Republic*, October 9, 1995, 16–25.

87. Eugene Garfield, "*Omni* Magazine Leads Upsurge in Mass-Audience Science Journalism," *Current Comments* 1979, 5–12.

88. Bruce V. Lewenstein, "The Arrogance of 'Pop Science,' " *Scientist*, July 13, 1987, 12. Lewenstein's essay generated several heated responses including an October 5, 1987, letter from Kathy Keeton claiming that, unlike other magazines, *Omni* continued to grow.

89. Jonathan R. Topham, "Historicizing Popular Science," *Isis* 2009, 100, 2: 310–18.

90. Ned Scharff, "Too Crowded Here? Why Not Fly into Space?" *Washington Star*, November 3, 1977. For background, chapter 5 of Michael Schaller, *Right Turn: American Life in the Reagan-Bush Era, 1980–1992* (New York: Oxford University Press, 2007).

91. NASA, *We Deliver*, 1983 brochure, included in John M. Logsdon et al., eds., *Exploring the Unknown: Selected Documents in the History of the U.S. Civil Space Program*, vol. 4, *Accessing Space* (Washington, DC: National Aeronautics and Space Administration, 1999), 423–26. The conservative space agenda is described in Andrew J. Butrica, *Single Stage to Orbit: Politics, Space Technology, and the Quest for Reusable Rocketry* (Baltimore: The Johns Hopkins University Press, 2003).

92. From Reagan's March 23, 1983, speech; for a good sampling of literature on SDI, see Donald R. Baucom, *The Origins of SDI, 1944–1983* (Lawrence: University of Kansas press, 1992), and Frances Fitzgerald, *Way Out There in the Blue: Reagan, Star Wars, and the End of the Cold War* (New York: Simon & Schuster, 2000).

93. *L5 News*, May 1976, 2; Peter J. Westwick, " 'Space-Strike Weapons' and the Soviet Response to SDI," *Diplomatic History* 2008, 32, 5: 955–79.

94. *L5 News*, July 1976, 5.

95. *L5 News*, July 1977, 7.

96. *L5 News*, June 1976, 5; *L5 News*, July 1977, 6.

97. Rebecca Slayton, "From Death Rays to Light Sabers," *Technology and Culture* 2011, 52, 1: 45–74.

98. Letter in the *L5 News*, September 1980, 18.

99. Michaud, *Reaching for the High Frontier*, 97.

100. William J. Broad, "Earthlings at Odds over Moon Treaty," *Science* 1979, 206, 4421: 915–16; Michaud, *Reaching for the High Frontier*, 94.

101. Letter from Jerry D. Campbell, *L5 News*, August 1985, 2.

102. "Satellites to Guide Air Traffic Backed," *New York Times*, April 2, 1979, D7; Gerard K. O'Neill, "Satellite-Based Vehicle Position Determining System," patent 4,359,733, issued by United States Patent and Trademark Office, November 16, 1982. Additional information comes from GKON and the GEO/NASM collection.

103. Like the much more sophisticated GPS system, also called Navstar, Geostar would have used geostationary satellites to receive signals from transceivers on the ground. Using calculations based on the different times the satellites

recorded the signals, a computer at a central ground station could calculate a position.

104. Jim Schefter, "Geostar," *Popular Science*, February 1984, 76–78, 130.

105. Gerard K. O'Neill, *The Technology Edge: Opportunities for America in World Competition* (New York: Simon and Schuster, 1983); July 20, 1981, memo from O'Neill to John Brockman, GKON.

106. For example, Ezra F. Vogel, *Japan as Number One: Lessons for America* (Cambridge, MA: Harvard University Press, 1979).

107. Freeman J. Dyson, "Gerard Kitchen O'Neill," *Physics Today*, February 1993, 97–98.

108. Marlise Simons, "A Final Turn-on Lifts Timothy Leary Off," *New York Times*, April 22, 1997, A1. Also, http://www.celestis.com/foundersFlight.asp (accessed August 2011).

Chapter 5: Could Small Be Beautiful?

1. Stewart Brand, ed., *Space Colonies: A CoEvolution Book* (San Francisco: Waller Press, 1977), 104.

2. Ibid., 90.

3. Letter from Drexler to Brand, undated, circa 1976, published in ibid., 90. Emphasis in original.

4. Quotes from Ed Regis, *Nano! Remaking the World Atom by Atom* (Boston: Little, Brown, 1995); Grant Fjermedal, *The Tomorrow Makers: A Brave New World of Living-Brain Machines* (New York: Macmillan Publishing Company, 1986), 175–76; Carolyn Henson, "How to Meet Space Entrepreneurs," *Future*, May 1979, 41; Fred Hapgood, "Tinytech," *Omni*, November 1986, 57; M. J. Wilcove, "I Have Seen the Future and It Is Tiny," *LA Reader*, November 3, 1989, 10; Mark Dowie, "The Last Industrial Revolution," *West*, January 1, 1989, 4–11, quote from 8. On scientist stereotypes, see Spencer R. Weart, "The Physicist as Mad Scientist," *Physics Today* 1988, 41, 6: 28–37.

5. There is no shortage of biographical material in print or online regarding Drexler; most of these items are short sketches of his background that appeared in conjunction with articles about nanotechnology. Two books by science writer Ed Regis also feature Drexler—*Great Mambo Chicken and the Transhuman Condition: Science Slightly over the Edge* (New York: Penguin Books, 1990), and *Nano!*. It should be noted that Drexler has disputed, in his correspondence with me, some of Regis's characterizations. Despite several inquiries, however, Drexler declined to give specific examples of Regis's errors. My biographical sketch of Drexler draws from all these sources as well as oral history interviews I did with him in December 2007 and Christine Peterson, his former spouse, in November 2006.

6. Regis, *Nano!* 87.

7. Eric Drexler, "Earth Day 1970, and my high road down to molecules," April 23, 2009, weblog posting, http://metamodern.com/2009/04/23/earth-day-1970-and-the-road-to-molecules/ (accessed December 2010).

8. K. Eric Drexler, December 11, 2007, interview with the author.

9. Regis, *Nano!*, 89

10. *Massachusetts Institute of Technology Bulletin*, September 1973, 108, 3: 191.

11. K. Eric Drexler, "Design of a High Performance Solar Sail System," MS thesis, Massachusetts Institute of Technology, May 1979.

12. Richard L. Garwin, "Solar Sailing: A Practical Method of Propulsion within the Solar System," *Jet Propulsion* 1958, 28: 188–89; also, Colin McInnes, "On the Crest of a Sunbeam," *New Scientist*, January 5, 1991, 31–33.

13. July 21, 1976, memo from A. M. Lovelace to James C. Fletcher, solar sail files, folder 13758, NASA/HO.

14. Peter J. Westwick, *Into the Black: JPL and the American Space Program, 1976–2004* (New Haven, CT: Yale University Press, 2007).

15. "Sailing to Halley's Comet," *Time*, March 14, 1977, 54.

16. European and Soviet missions, however, did successfully study the comet in 1986; John M. Logsdon, "Missing Halley's Comet: The Politics of Big Science," *Isis* 1989, 80, 2: 254–80.

17. Jerome Wright, *Solar Sailing* (New York: Gordon and Breach, 1992). NASA and the Japanese Aerospace Exploration Agency have since deployed and tested solar sail prototypes. In June 2005, the Planetary Society, a publicly funded nongovernmental group, launched a solar sail project, but the Russian Volna rocket it was on failed to reach orbit.

18. K. Eric Drexler, MIT Space Systems Laboratory Report 5-79, 1979.

19. Eric Drexler, "Lightsail Update," *L5 News*, November 1977, 1.

20. The application was submitted May 5, 1980; Kim E. Drexler, "Solar Sail," United States Patent 4,614,319, issued by United States Patent and Trademark Office, September 30, 1986.

21. For example, Eric Drexler, "Sailing on Sunlight May Give Space Travel a Second Wind," *Smithsonian*, February 1982, 52–61.

22. K. Eric Drexler, "Exploring Future Technologies," in *The Reality Club*, edited by John Brockman (New York: Lynx Books, 1988), 129–50.

23. Ibid., 134–35.

24. K. Eric Drexler, December 11, 2007, interview with the author.

25. Drexler, "Exploring Future Technologies," 150.

26. Gerard K. O'Neill, *The High Frontier: Human Colonies in Space* (New York: William Morrow and Company, 1977), 11.

27. Drexler, "Earth Day 1970."

28. Regis, *Nano!* 45.

29. Hapgood, "Tinytech," 56–62, 202.

30. Regis, *Nano!*, 47–48.

31. Ibid.

32. Ibid.

33. Simon Schaffer, "Enlightened Automata," in *The Sciences in Enlightened Europe*, edited by William Clark, Jan Golinski, and Simon Schaffer (Chicago: University of Chicago Press, 1999), 126–65.

34. Jessica Riskin, "The Defecating Duck; or, The Ambiguous Origins of Artificial Life," *Critical Inquiry* 2003, 29, 4: 599–633, quote from 622.

35. This anecdote appears often in discussions about self-replication such as

Moshe Sipper and James A. Reggia, "Go Forth and Replicate," *Scientific American*, August 2001, 34–43.

36. Von Neumann's ideas were published after his death in 1957 as *The Computer and the Brain* (New Haven, CT: Yale University Press, 1958); his papers on self-replication were published as *Theory of Self-Reproducing Automata*, edited by A. W. Burks (Urbana: University of Illinois Press, 1966).

37. Moshe Sipper, "Fifty Years of Research on Self-Replication: An Overview," *Artificial Life* 1998, 4, 3: 237–57; Robert A. Freitas, Jr., and Ralph C. Merkle, *Kinematic Self-Replicating Machines* (Georgetown, TX: Landes Bioscience, 2004).

38. L. S. Penrose, "Self-Reproducing Machines," *Scientific American*, June 1959, 105–17, quote from front piece.

39. Homer Jacobson, "On Models of Reproduction," *American Scientist* 1958, 46, 3: 255–84.

40. Edward F. Moore, "Artificial Living Plants," *Scientific American*, October 1956, 118–26.

41. Freeman J. Dyson, March 8, 2008, e-mail to the author.

42. Moore, "Artificial Living Plants," 124.

43. Excerpts of Frosch's September 14, 1979, speech appeared in *L5 News*, January 1980, 12.

44. Robert A. Freitas, Jr., and William P. Gilbreath, eds., *Advanced Automation for Space Missions* (Washington, DC: National Aeronautics and Space Administration, 1982).

45. Stanisław Lem, *The Invincible* (New York: Seabury Press, 1973 [English edition]; originally published 1964).

46. Background on Peterson comes from Regis, *Nano!* as well as my November 28, 2006, interview with her.

47. Lewis Branscomb, "Microscience and Basic Research," *Physics Today*, November 1979, 112; also, James A. Krumhansl and Yoh-Han Pao, "Microscience: An Overview," *Physics Today*, November 1979, 25–33.

48. Over four hundred articles cite "Plenty of Room" as of this writing, with the majority appearing after 1995. Christopher Toumey, March 10, 2011, personal communication.

49. Richard P. Feynman, "There's Plenty of Room at the Bottom," *Engineering and Science*, February 1960, 22–36, quote from 34.

50. Regis, *Nano!* 152–54; also, Andreas Junk and Falk Riess, "From an Idea to a Vision: There's Plenty of Room at the Bottom," *American Journal of Physics* 2006, 74, 9: 825–30, and Colin Milburn, *Nanovision: Engineering the Future* (Durham, NC: Duke University Press, 2008).

51. Feynman, "There's Plenty of Room at the Bottom," 34.

52. Arthur R. Von Hippel, "Molecular Engineering," *Science* 1956, 123, 3191: 315–17.

53. L. E. Cross and R. E. Newnham, "History of Ferroelectrics," in *High Technology Ceramics: Past, Present, and Future*, edited by W. D. Kingery (Westerville, OH: American Ceramic Society, 1987).

54. Feynman, "There's Plenty of Room at the Bottom," 36.

55. Arthur R. Von Hippel, "Molecular Designing of Materials," *Science* 1962, 138, 3537: 91–108.

56. Dennis Overbye, "Arthur R. Kantrowitz, Whose Wide-Ranging Research Had Many Applications, Is Dead at 95," *New York Times*, December 8, 2008.

57. Arthur Kantrowitz, "Proposal for an Institution for Scientific Judgment," *Science* 1967, 156, 3776: 763–64.

58. John Noble Wilford, "Experts Back Plans for 'Science Court,'" *New York Times*, September 22, 1976, 11; Dorothy Nelkin, "Thoughts on the Proposed Science Court," *Newsletter on Science, Technology, & Human Values* January 1977, 18: 20–31.

59. K. Eric Drexler, December 11, 2007, interview with the author.

60. Undated, likely early 1981, letter from Drexler to Morrison; box 1, ARK/DC.

61. K. Eric Drexler, September 6, 2010, e-mail to the author.

62. K. Eric Drexler, "Molecular Engineering: An Approach to the Development of General Capabilities for Molecular Manipulation," *Proceedings of the National Academy of Science* 1981, 78, 9: 5275–78.

63. Ibid., 5275.

64. Ibid., 5276.

65. K. Eric Drexler, September 8, 2008 e-mail to the author.

66. Drexler, "Molecular Engineering," 5275.

67. Emily Martin, *The Woman in the Body: A Cultural Analysis of Reproduction* (Boston: Beacon Press, 1987).

68. Drexler, "Molecular Engineering," 5276–77.

69. Ibid., 5278.

70. April 5, 1981, letter from Drexler to Kantrowitz, box 1, ARK/DC.

71. Freeman J. Dyson, *PNAS* referee report, circa March 1981, box 1, ARK/DC.

72. Philip Morrison, *PNAS* referee report, circa March 1981, box 1, ARK/DC.

73. February 16, 1981, letter from Drexler to Kantrowitz; box 1, ARK/DC.

74. Regis, *Nano!* 110.

75. Alex Roland and Philip Shiman, *Strategic Computing: DARPA and the Quest for Machine Intelligence* (Cambridge, MA: MIT Press, 2002).

76. Eric Drexler, "Mightier Machines from Tiny Atoms May Someday Grow," *Smithsonian*, November 1982, 145–54, emphasis mine.

77. William F. DeGrado, F. J. Kezdy, and E. T. Kaiser, "Design, Synthesis, and Characterization of a Cytotoxic Peptide with Melittin-Like Activity," *Journal of the American Chemical Society* 1981, 103, 3: 679–81.

78. Carl O. Pabo, "Molecular Technology: Designing Proteins and Peptides," *Nature* 1983, 301, 7330: 200.

79. Robert Reinhold, "Bacteria Tycoons Start a Real Growth Industry," *New York Times*, February 3, 1980, E8.

80. "Foresight Seminar of the Future of Biotechnology," December 15, 1982, http://www.altfutures.org/for_sem_all (accessed January 2011).

81. Kevin Ulmer, February 14, 2011, e-mail to the author.

82. Hyungsub Choi and Cyrus C. M. Mody, "The Long History of Molecular

Electronics: Microelectronics Origins of Nanotechnology," *Social Studies of Science* 2009, 39, 1: 11–50.

83. For example, Arieh Aviram and Mark A. Ratner, "Molecular Rectifiers," *Chemical Physics Letters* 1974, 29, 2: 277–83.

84. Choi and Mody, "Molecular Electronics," 30.

85. Stephanie Yanchinski, "And now—the biochip," *New Scientist*, January 14, 1982, 68–71; Natalie Angier, "The Organic Computer," *Discover*, May 1982, 76–79; Arthur L. Robinson, "Nanocomputers from Organic Molecules," *Science* 1983, 220, 4600: 940–942.

86. "Foresight Seminar of the Future of Biotechnology," December 15, 1982.

87. Kevin Ulmer, "Protein Engineering," *Science* 1983, 219, 4585: 666–71.

88. Ibid., 670.

89. Eric Drexler, "The Future by Design," 1983, draft book manuscript, KH/SA.

90. Ibid., 87, 90.

91. J. Baldwin, "One Highly Evolved Toolbox," *CoEvolution Quarterly*, Spring 1975, 80–85. Also, see Andrew G. Kirk, *Counterculture Green: The Whole Earth Catalog and American Environmentalism* (Lawrence: University of Kansas Press, 2007), 64.

92. Described by J. Baldwin in *The Next Whole Earth Catalog* (New York: Rand McNally, 1980), 138.

93. Drexler, "The Future by Design," 142.

94. Undated (c. 1983) letter from Eric Drexler to Keith Henson, KH-AL/SA.

95. Conrad Schneiker, November 9, 2007, interview with the author. More background on Schneiker appears in articles by Christopher Toumey including "The Man Who Understood the Feynman Machine," *Nature Nanotechnology* 2007, 2, 1: 9–10, and "Reading Feynman into Nanotechnology," *Techné* 2008, 12, 3: 133–68.

96. September 12, 1982, letter and comments from Schneiker to Drexler, CWS.

97. David C. Brock, August 30, 2011, personal communication to the author.

98. Gerd Binnig and Heinrich Rohrer, "Scanning Tunneling Microscopy: From Birth to Adolescence," *Reviews of Modern Physics* 1987, 59, 3: 615–25; Christoph Gerber and Hans Peter Lang, "How the Doors to the Nanoworld Were Opened," *Nature Nanotechnology* 2006, 1, 10: 3–5.

99. Gerber and Lang, "Doors to the Nanoworld."

100. Cyrus C. M. Mody, *Instrumental Community: Probe Microscopy and the Path to Nanotechnology* (Cambridge, MA: MIT Press, 2011).

101. Conrad Schneiker, "The Modified Scanning Tunneling Microscope as a Nanometer Scale Machine Tool and Multimode Interface for Precision Assembly, Manipulation, Analysis, and Control of Solid State Atomic Systems," unpublished manuscript, February 26, 1985, CWS.

102. Reflected in a series of letters from 1985 in Schneiker's personal papers, copies in the author's possession.

103. October 16, 1985, letter from Hansma to Feynman, CWS.

104. Gerd Binnig and Heinrich Rohrer, December 8, 1986, Nobel Lecture, reprinted as "Scanning Tunneling Microscopy: From Birth to Adolescence," *Reviews of Modern Physics* 1987, 59, 3: 615–25.

105. R. S. Becker, J. A. Golovchenko, and B. S. Swarzentruber, "Atomic-Scale Surface Modifications Using a Tunneling Microscope," *Nature* 1987, 325, 6103: 419–21.

106. Stuart Hameroff et al., "NanoTechnology Workstation Based on Scanning Tunneling/Optical Microscopy: Applications to Molecular Devices," October 6, 1986, manuscript, CWS; later published in *Molecular Electronic Devices*, edited by F. L. Carter, R. E. Siatkowski, and H. Wohltjen (Amsterdam: Elsevier Science Publishers, 1988), 69–90.

107. Norio Taniguchi, "On the Basic Concept of 'Nano-Technology,'" in *Proceedings of the International Conference of Production Engineering* (Tokyo: Japan Society of Precision Engineering, 1974).

108. Conrad Schneiker, November 9, 2007, interview with the author. Also, see comments on Schneiker's website, http://www.athenalab.com/NanoScam.htm (accessed August 2011).

109. Quote is from a talk Drexler gave May 25, 1985, at the Lake Tahoe Life Extension Festival; a transcript of the talk appears as Eric Drexler, "Molecular Technology and Cell Repair Machines," *Cryonics*, published in two parts in December 1985 and January 1986, issues 65 and 66.

110. Letter from Schneiker published in *Cryonics*, December 1985, 65: 9–12; also, Conrad Schneiker, November 9, 2007, interview with the author.

111. Conrad Schneiker, November 9, 2007, interview with the author.

112. Conrad Schneiker, August 23, 2011, e-mail to the author. For example, in 1988, he and an IBM researcher applied to patent a "distance-controlled tunneling transducer." This was issued by the United States Patent and Trademark Office, August 27, 1991, as patent #5,043,577.

113. Conrad Schneiker, November 9, 2007, interview with the author.

Chapter 6: California Dreaming

1. This comment was made by Richard A. L. Jones, a British physicist, Fellow of the Royal Society, and writer of a popular book and blog on nanotechnology in a review titled "Hollow Centre," *Nature* 2006, 440, 7087: 995.

2. This and many other similar phrases accompanied media coverage of nanotech in the early twenty-first century. This particular example comes from a special section called "The Nano Age," which appeared in the April 17, 2006, issue of *Fortune*.

3. Conrad Schneiker, November 9, 2007, interview with the author.

4. Henry Etzkowitz, *The Triple Helix: University-Industry-Government Relations in Action* (New York: Routledge, 2008).

5. "Drexler on Drexlerians," December 15, 2003, post on Howard Lovy's NanoBot blog, http://nanobot.blogspot.com/2003/12/drexler-on-drexlerians.html (accessed January 2011).

6. Lowell Ponte, "Dawn of The 'Tiny Tech' Age," *Reader's Digest*, November 1990, 25.

7. Eric Drexler, August 2007 interview with Michael Lounsbury, transcript in author's possession.

8. March 5, 1985, letter from Eric Drexler to Charles "Ed" Tandy, published in *Cryonics*, May 1985, 58: 5–6.

9. Eric Drexler, January 25, 2011, e-mail to the author. K. Eric Drexler, *Engines of Creation: The Coming Era of Nanotechnology* (New York: Anchor Books, 1986); hereafter *Engines*. Anchor Books was an imprint of Doubleday.

10. *Engines*, 182; K. Eric Drexler, March 14, 2011, e-mail to the author.

11. Terence Monmaney, "Nanomachines to Our Rescue," *New York Times*, August 10, 1986, BR8.

12. J. Baldwin, "Engines of Creation," *Whole Earth Review*, Winter 1986, 83.

13. "Nanotechnology Press Kit," prepared by the MIT Nanotechnology Study Group, July 1987, JCB.

14. "David R. Forrest, "Nanotechnology," April 1986 manuscript, CWS.

15. Martin Brooks, *Nanotech News*, October 1987, CWS.

16. Ed Regis, *Nano! Remaking the World Atom by Atom* (Boston: Little, Brown, 1995), 140.

17. "Purposes of the Foresight Institute," July 11, 1986, draft, JCB.

18. Eric Drexler, December 11, 2007, interview with the author.

19. James C. Bennett, July 19, 2007, interview with the author.

20. Nadrian C. Seeman, March 16, 2011, e-mail to the author.

21. June 17, 1987, letter from Peterson to James Bennett, JCB. Christine Peterson, November 30, 2006, e-mail to the author.

22. Drexler quoted in M. J. Wilcove, "Nanotechnology: Schmoo or Gray Goo," April 1988, draft manuscript for *L.A. Weekly*, JCB.

23. Fred Hapgood, "Tinytech," *Omni*, November 1986, 56–62, 202. A retrospective view is Eric Drexler, "The promise that launched the field of nanotechnology," December 15, 2009, blog posting, http://metamodern.com/2009/12/15/when-a-million-readers-first-encountered-nanotechnology/ (accessed January 30, 2011).

24. Hapgood, "Tinytech," 62.

25. A. K. Dewdney, "Nanotechnology: Wherein Molecular Computers Control Tiny Circulatory Submarines," *Scientific American*, January 1988, 100–103.

26. Karen Breslau, "The Doctor That Floats in Your Bloodstream," *New York Times Magazine*, June 11, 2000, 101–102.

27. *Foresight Update* #3, April 30, 1988, 2.

28. Wil McCarthy, "Nanotechnology: Abuses of and Replacements For," *Bulletin of the Science Fiction Writers of America*, Fall 2001, 20–23. My thanks to Colin Milburn for pointing this out.

29. Based on an online list complied by Anthony Napier; although the web link has disappeared, it can be recovered via the Internet Archive. Also, see Brooks Landon, "Less Is More: Much Less Is Much More; The Insistent Allure of Nanotechnology Narratives in Science Fiction Literature," in *Nanoculture: Implications of the New Technoscience*, edited by N. Katherine Hayles (Portland, OR: Intellect Books, 2005), 131–46.

30. Howard E. McCurdy, *Space and the American Imagination* (Washington, DC: Smithsonian Institution Press, 1997).

31. Milburn, *Nanovision*, quote from p. 13.

32. Daniel Patrick Thurs, "Tiny Tech, Transcendent Tech: Nanotechnology,

Science Fiction, and the Limits of Modern Science Talk," *Science Communication* 2007, 29, 1: 65–95.

33. Fred Gardner, "Nano and the Professor," *Interview*, January 1989, 92–93, 110.

34. *Foresight Update* #2, November 15, 1987, 1.

35. For example, November 2, 1987, letter from Michael Korns, a former IBM executive, to Eric Drexler, JCB.

36. Regis, *Nano!* 189–90. Additional information on Foresight fund-raising comes from documents shared with the author by James Bennett.

37. "The Great Migration," in *The Papers of Robert H. Goddard*, edited by Robert Goddard and G. E. Pendray (New York, 1970), vol. 3, 1611–12.

38. Manfred E. Clynes and Nathan S. Kline, "Cyborgs and Space," *Astronautics* 1960, 26–27, 74–75; "Spaceman Is Seen as Man-Machine," *New York Times*, May 22, 1960, 31.

39. Gerald Feinberg, "Physics and Life Prolongation," *Physics Today* 1966, 19:45–48. Feinberg went on to write several books arguing for society's need to pursue bold technology projects, including *The Prometheus Project: Mankind's Search for Long-Range Goals* (New York: Doubleday, 1968) and *Consequences of Growth: The Prospects for a Limitless Future* (New York: Seabury Press, 1977).

40. Robert Ettigner, *The Prospect of Immortality*, self-published in 1962 (Doubleday edition in 1964); a review of it is D. E. Goldman, "American Way of Life?" *Science* 1964, 145: 475–76.

41. David Larsen, "Cancer Victim's Body Frozen for Future Revival Experiment," *Los Angeles Times*, January 19, 1967. The story was also covered in the *New York Times* and described in a self-published book from 1968 called *We Froze the First Man*.

42. William H. Honan, "The Futurists Take over the Jules Verne Business," *New York Times*, April 9, 1967, M243.

43. Ken Lubas, "Cryonics Society's Facility for Frozen Death to Open," *Los Angeles Times*, April 20, 1969, B1.

44. "Instructions for the Induction of Solid-State Hypothermia in Humans," http://www.lifepact.com/mm/mrm001.htm (accessed July 2010).

45. Described on Alcor's website: http://www.alcor.org/AboutAlcor/nameori gin.html (accessed July 2010).

46. For example, Keith Henson, "When We Wake Up," *Cryonics*, October 1986, 29–31. Also January 1985 letter in *Cryonics*.

47. "Molecular Engineering," *Cryonics*, April 1984, 5, and online, http://www. alcor.org/cryonics/cryonics8404.txt (accessed July 2010).

48. For example, Drexler gave a talk in May 1985 at the Lake Tahoe Life Extension Festival called "Molecular Technology and Cell Repair Machines," reprinted in the December 1985 issue of *Cryonics*, http://www.alcor.org/cryonics/cryonics8512.txt (accessed July 2010).

49. Michael Cieply, "They Freeze Death If Not Taxes," *Los Angeles Times*, September 9, 1990; also, see http://www.alcor.org/AboutAlcor/membershipstats .html (accessed August 2011).

50. Tim Larimer, "The Next Ice Age," *West* (magazine supplement to the *San Jose Mercury News*), December 9, 1990, 17–26.

51. Although Alcor maintains anonymity for its "patients," it is possible to connect the "de-animation" described in "The Transport of Patient A-1312" in *Cryonics* #139, February 1992, written by Keith Henson, to Salin.

52. Theodor H. Nelson, "A File Structure for the Complex, the Changing, and the Indeterminate," in *Association for Computing Machinery: Proceedings of the 20th National Conference*, edited by Lewis Winner (New York: Association for Computing Machinery, 1965).

53. Vannevar Bush, "As We May Think," *Atlantic Monthly*, July 1945, 101–8; an illustrated abridged version appeared in *Life* in September 1945.

54. Daniel Rosenberg, "Electronic Memory," in *Histories of the Future*, edited by Daniel Rosenberg and Susan Harding (Durham, NC: Duke University Press, 2005), 125–51; a recollection by World Wide Web "father" Tim Berners-Lee credits both Bush and Nelson's ideas; "www: Past, Present, and Future," *Computer* 1996, 29, 10: 69–77.

55. From the cover of Theodor Holm Nelson, *Computer Lib/Dream Machines* (self-published, 1974).

56. Gary Wolf, "The Curse of Xanadu," *Wired*, June 1995 (http://www.wired.com/wired/archive/3.06/xanadu.html) presents a version of Xanadu's history, one that many of its participants disagreed with; see responses to Wolf's article, http://www.wired.com/wired/archive/3.09/rants.html (accessed July 2011). Also, see Stuart Moulthrop, "You Say You Want a Revolution: Hypertext and the Laws of Media," *Postmodern Culture*, May 1991, http://muse.jhu.edu/journals/postmodern_culture/summary/v001/1.3moulthrop.html (accessed July 2011).

57. Quotes from *Engines*, 224; chapters 13 and 14 give a fuller explication.

58. For orchestrated misinformation campaigns, see Naomi Oreskes and Erik M. Conway, *Merchants of Doubt: How a Handful of Scientists Obscured the Truth on Issues from Tobacco Smoke to Global Warming* (New York: Bloomsbury Press, 2010).

59. G. Pascal Zachary, "Theocracy of Hackers Rules Autodesk," *Wall Street Journal*, May 28, 1992, 1.

60. From Walker's online memoir, http://www.fourmilab.ch/autofile/www/autofile.html (accessed January 2011).

61. James C. Bennett, July 19, 2007, interview with the author.

62. *Foresight Update* #11, March 15, 1990, 2.

63. Christine L. Peterson, November 28, 2006, interview with the author.

64. Mark S. Miller, May 11, 2010, e-mail to the author.

65. Charles Alexander, Adam Zagorin, and Gisela Bolt, "The New Economy," *Time*, May 30, 1983, 62–70.

66. For example, James Gleick, *Chaos: Making a New Science* (New York: Penguin Books, 1987), and M. Mitchell Waldrop, *Complexity: The Emerging Science at the Edge of Order and Chaos* (New York: Simon & Schuster, 1992).

67. John Horgan, "From Complexity to Perplexity," *Scientific American*, June 1995, 104–9.

68. Robert Pool, "Strange Bedfellows," *Science* 1989, 245, 4919: 700–703. The possible linkages between scientists' interests and larger historical trends was

cautiously confirmed by anecdotal evidence from M. Mitchell Waldrop, who wrote a 1992 book on complexity studies; February 24, 2011, correspondence with the author.

69. Mark Sinker, "Nanotechnology: This Year's Chaos," *Sunday Correspondent*, June 3, 1990, 47.

70. The phrase comes from a September 22, 1989, talk by Jay Ogilvy, folder 4, box 66, SB/SA.

71. Office of Technology Assessment, *Technology and the American Economic Transition: Choices for the Future* (Washington, DC, 1988), 16.

72. Turner, *From Counterculture to Cyberculture*, chapter 6. More background on GBN comes from Joel Garreau, "Conspiracy of Heretics," *Wired*, November 1994, 98–106, 153–58.

73. January 2, 1988, notes from GBN planning meeting, folder 2, box 68, SB/SA.

74. January 6, 1988, memo from Peter Schwartz, folder 1, box 68, SB/SA.

75. 1988 GBN promotional brochure, folder 1, box 66, SB/SA.

76. Garreau, "Conspiracy of Heretics."

77. Turner, *From Counterculture to Cyberculture*, 191.

78. Peter Schwartz, *The Art of the Long View* (New York: Doubleday/Currency, 1991), 92; Garreau, "Conspiracy of Heretics."

79. Undated (1988) planning document, folder 2, box 68, SB/SA.

80. December 1988 issue of GBN publication the *Deeper News* as well as the "1990 Scenario Book," folder 1, box 66 and box 75, folder 7, both SB/SA.

81. Published as Stewart Brand, *The Media Lab: Inventing the Future at MIT* (New York: Penguin, 1988).

82. Stewart Brand, "Mothers of Invention," *Omni*, August 1986, 18.

83. *GBN Book Club*, November 1988, folder 1, box 75, SB/SA, and http://www.gbn.com/BookClubSelectionDisplayServlet.srv?si=90 (accessed February 2011).

84. May 22, 1992, letter from GBN to Paul Falcone and September 22, 1989, presentation by Ogilvy, folder 3, box 79, and folder 4, box 66, both SB/SA.

85. "International Interest in Nanotechnology," *Foresight Update* #8, March 15, 1990; September 12, 1988, note from Stewart Brand, folder 3, box 66, SB/SA.

86. From Brand's foreword to K. Eric Drexler, Chris Peterson, and Gayle Pergamit, *Unbounding the Future: The Nanotechnology Revolution* (New York: Quill, 1991), 7.

87. Grant Fjermedal, *The Tomorrow Makers: A Brave New World of Living-Brain Machines* (New York: Macmillan Publishing Company, 1986).

88. Diary notes from February 7 and April 5, 1987, folder 10, box 20, SB/SA.

89. The reference is to the Homebrew Computer Club, the legendary hobbyist group; "Minutes of the 4th Seattle NSG Meeting," November 24, 1987, CWS.

90. From introduction to *Nanocon Proceedings*, July 1989 (self-published by Nanocon), CWS, and online, http://www.halcyon.com/nanojbl/NanoConProc/nanocon1.html (accessed July 2011).

91. Undated (late 1988) e-mail from Brand to Eric Drexler and Christine Peterson, folder 1, box 66, SB/SA.

92. "Foresight Conference Publicity Plan," July 29, 1989, FI.

93. Ibid.

94. "Stanford Hosts Foresight Conference," November 13, 1989, press release, FI.

95. Philip Elmer-DeWitt," The Incredible Shrinking Machine," *Time*, November 20, 1989, 108–9.

96. "The Invisible Factory," *Economist*, December 9, 1989, 91–92.

97. Brett Duval Fromson, "Where the Next Fortunes Will Be Made," *Fortune*, December 5, 1988, 185–89.

98. Garfinkel and Drexler's exchange appears in the Summer 1990 issue of the *Whole Earth Review*, 104–13; emphasis in original.

99. Steven Levy, "Maximizing the Understanding," *Whole Earth Review*, Summer 1990, 114.

100. K. Eric Drexler, "Machines of Inner Space," in *Encyclopedia Britannica Yearbook of Science and the Future 1990*, edited by David Calhoun (Chicago: University of Chicago Press, 1989), 162–77.

101. Foresight's own definitions varied over time; this is taken from *Foresight Update* #17, December 15, 1993, 4.

102. "Senior Associates Program Description," undated but likely 1992, KH-AL/SA.

103. "IMM to Fund Molecular Manufacturing," *Foresight Update* #12, August 1, 1991, 1.

104. *Foresight Update* #17, December 15, 1993, 5.

105. Described in *Foresight Update* #28, March 30, 1997, 2, and http://www.foresight.org/about/MatchingGrant.html (accessed March 2011)

106. Junk and Riess, "From an Idea to a Vision," 827–28.

107. *Foresight Update* #24, April 15, 1996, 1. As of July 2012, the prize is unclaimed.

108. James R. Von Ehr II, January 24, 2011, interview with David C. Brock, CHF.

109. "First Nanotechnology Development Company Formed," *Foresight Update* #29, June 30, 1997, 3.

110. "New Technologies for a Sustainable World," June 26, 1992, hearing before the Subcommittee on Science, Technology, and Space, United States Senate, 102nd Congress.

111. Described in Regis, *Nano!* 217–20, 236–39, 249–52.

112. Ibid., 237.

113. Gary Stix, "Waiting for Breakthroughs," *Scientific American* 1996, 274, 4: 96.

114. Robert Langreth, "Molecular Marvels," *Popular Science*, May 1993, 91–94, 110–11.

115. K. Eric Drexler, *Nanosystems: Molecular Machinery, Manufacturing, and Computation* (New York: John Wiley & Sons, 1992), 445.

116. Philip Ball, "Small Problems," *Nature* 1993, 362, 6416: 123.

117. Quote from P. L. Anelli, N. Spencer, and J. F. Stoddart, "A Molecular Shuttle," *Journal of the American Chemical Society* 1991, 113, 13: 5131–33; also, see Peter R. Ashton et al., "Molecular Trains: The Self-Assembly and Dynamic

Properties of Two New Catenanes," *Angewandte Chemie International Edition in English* 1991, 30: 1042–45.

118. Ball, "Small Problems."

119. For example, David E. H. Jones, "Technical Boundless Optimism," *Nature* 1995, 374, 6525: 835–37; David Voss, "Moses of the Nanoworld," *Technology Review* 1999: 60–62.

120. "First Nanosystems Course Now in Progress at USC," *Foresight Update* #18, April 15, 1994, 1.

121. Gregg Herken, "The Flying Crowbar," *Air & Space Magazine*, April/May 1990, 28–34.

122. See, for example, November 22, 2004, oral history interview with Martin Hellman, done by Jeffrey Yost, OH 375, Charles Babbage Institute, University of Minnesota, Minneapolis, MN; and Whitfield Diffie, "The First Ten Years of Public-Key Cryptography," *Proceedings of the IEEE* 1988, 76, 5: 560–77.

123. Christine Blackman, "Stanford Encryption Pioneer Who Risked Career Becomes Hamming Medalist," *Stanford Report*, February 10, 2010, http://news.stanford.edu/news/2010/february8/hellman-encryption-medal-021010.html (accessed March 2011).

124. Ralph C. Merkle, "Computational Nanotechnology," *Nanotechnology* 1991, 2, 3: 134–41.

125. Eric Winsberg, *Science in the Age of Computer Simulation* (Chicago: University of Chicago Press, 2010).

126. Ann Johnson, "Institutions for Simulations: The Case of Computational Nanotechnology," *Science Studies* 2006, 19, 1: 35–51; Ann Johnson, "Modeling Molecules: Computational Nanotechnology as Knowledge Community," *Perspectives on Science* 2009, 17, 2: 144–73.

127. Merkle, "Computational Nanotechnology"; also see Ralph C. Merkle, "Computational Nanotechnology at Xerox PARC," *Foresight Update* #11, March 15, 1991, 2.

128. Quotes from Mary Eisenhart, "Nanotechnology: Separating Myth from Reality (an Interview with Eric Drexler)," *Microtimes*, October 26, 1992, 130–34, 79–82, 220.

129. Meyya Meyyappan, July 24, 2009, interview with the author; Scott Hubbard, September 24, 2009, interview with the author.

130. Al Globus, September 24, 2009, interview with the author; Richard Jaffe, September 25, 2009, interview with the author; Deepak Srivastava, September 23, 2009, interview with the author.

131. Al Globus et al., "NASA Applications of Molecular Nanotechnology," *Journal of the British Interplanetary Society* 1998, 51: 145–52.

132. Al Globus et al., "Machine Phase Fullerene Nanotechnology," *Nanotechnology* 1998, 9, 3: 192–99.

133. Glenn E. Bugos, *Atmosphere of Freedom: 70 Years at the NASA Ames Research Center* (Washington, DC: National Aeronautics and Space Administration History Office, 2010), 141–51, describes the rise and fall of NASA's nanotechnology program.

134. Deepak Srivastava, September 23, 2009, interview with the author.

135. My emphasis; quote from David Rotman, "Will the Real Nanotech Please Stand Up?" *Technology Review*, March/April 1999, 46–53.

Chapter 7: Confirmation, Benediction, and Inquisition

1. The bill's cosponsor Senator George Allen, quoted in Juliana Gruenwald, "D.C., Nano Union Now Put to the Test," *smalltimes*, January–February 2004, 8–9.

2. The "father" and "godfather" appellations appear in many places including "Nanotechnology's Unhappy Father," *Economist*, March 11, 2004, http://www.economist.com/node/2477051 (accessed April 2011).

3. Remark by physicist Richard A. L. Jones in 2006, quoted in Arie Rip and Marloes Van Ameron, "Emerging *De Facto* Agendas Surrounding Nanotechnology: Two Cases Full of Contingencies, Lock-Outs, and Lock-Ins," in *Governing Future Technologies*, edited by Mario Kaiser et al. (New York: Springer, 2010), 131–55.

4. Eric Drexler, "Mightier Machines from Tiny Atoms May Someday Grow," *Smithsonian*, August 1982, 145–54.

5. Karl Popper, *Conjectures and Refutations: The Growth of Scientific Knowledge* (New York: Routledge, 2002 [1963]), 48.

6. Thomas Kuhn, *The Structure of Scientific Revolutions* (Chicago: University of Chicago Press, 1962).

7. The term "crypto-history" comes from Michael Schiffer, *The Portable Radio in American Life* (Tucson: University of Arizona Press, 1991). Schiffer used the term to describe history deployed to serve the needs of certain actors and interests while obscuring other actors and stories; specifically, he was responding to the oft-unchallenged and erroneous claim that Japanese firms introduced the portable transistor radio to American markets.

8. Tim Wu, *The Master Switch: The Rise and Fall of Information Empires* (New York: Alfred A. Knopf, 2010), 149–54.

9. For example, Ivan Amato, *Nanotechnology: Shaping the World Atom by Atom* (Washington, DC: National Science and Technology Council, 1999).

10. Adam Keiper, "The Nanotech Schism," *New Atlantis*, Winter 2004, http://www.thenewatlantis.com/publications/the-nanotech-schism (accessed January 2011); also, David Berube and J. D. Shipman, "Denialism: Drexler Vs. Roco," *IEEE Technology and Society Magazine*, Winter 2004, 22–26.

11. August 28, 1993, letter from Richard Smalley to Nai-Teng Yu, folder 1, box 3, RS/RU.

12. "Bucky Balls, Fullerenes, and the Future: An Oral History with Professor Richard E. Smalley," January 22, 2000, interview with Smalley by Robbie Davis-Floyd and Kenneth J. Cox, http://www.davis-floyd.com/ShowPage.asp?id=155 (accessed March 2011).

13. Background on Smalley's research comes from a number of sources including the entry for Smalley in the *New Dictionary of Scientific Biography* (accessed January 2011) and Cyrus C. M. Mody, *Institutions as Stepping-Stones: Rick*

Smalley and the Commercialization of Nanotubes (Philadelphia: Chemical Heritage Foundation, 2010).

14. Daniel Koshland, Jr., "Molecule of the Year," *Science* 1991, 254, 5039: 1705.

15. Cyrus C. M. Mody, "Introduction," *Perspectives on Science* 2009, 17, 2: 111–22; Mody, August 31, 2011, personal communication to the author.

16. Malcolm W. Browne, "It's a Nanometer in Diameter, and Now It Can Be Made in Bulk," *New York Times*, June 22, 1993, C11.

17. *Carbon Nanoscale Laboratory*, Rice University promotional brochure (undated but likely mid-1990s), folder 8, box 3, RS/RU.

18. From "Bucky Balls, Fullerenes, and the Future."

19. January 21, 1993, letter from Richard Smalley to Michael M. Carroll and James L. Kinsey, folder 3, box 3, RS/RU.

20. Todd Ackerman in *Houston Chronicle*, November 11, 1993, 38A.

21. For example, an October 5, 1993, memo from Jackie Bourne to Richard Smalley notes the mailing of copies of *Unbounding the Future*, folder 7, box 21, RS/RU.

22. Arthur C. Clarke, *The Fountains of Paradise* (New York: Harcourt Brace Jovanovich, 1979).

23. August 24, 1996, e-mail from Smalley to Gustavo Scuseria, box 59, folder 7, RS/CHF.

24. K. Eric Drexler and John S. Foster, "Synthetic Tips," *Nature* 1990, 343, 6259: 600.

25. D. M. Eigler and E. K. Schweizer, "Positioning Single Atoms with a Scanning Tunnelling Microscope," *Nature* 1990, 344, 6266: 524–26; Malcolm W. Browne, "2 Researchers Spell 'IBM,' Atom by Atom." *New York Times*, April 5, 1990, B11.

26. Christopher Toumey, "35 Atoms That Changed the Nanoworld," *Nature Nanotechnology* 2010, 5, 4: 239–41.

27. For example, Semiconductor Industry Association, *The National Technology Roadmap for Semiconductors*, San Jose, CA, 1997.

28. For one perspective, see chapter 1 of Clayton M. Christensen, *The Innovator's Dilemma: When New Technologies Cause Great Firms to Fail* (Cambridge, MA: Harvard Business Press, 1997).

29. W. Patrick McCray, "From Lab to iPod: A Story of Discovery and Commercialization in the Post–Cold War Era," *Technology and Culture* 2009, 50, 1: 58–81.

30. Described in numerous publications including materials at the IBM Web site: http://www.research.ibm.com/research/gmr.html and http://archive.sciencewatch.com/interviews/stuart_parkin1.htm (accessed April 2011). The quote is from David D. Awschalom, April 2006 interviews with the author.

31. Raju Narisetti, "IBM Unveils Powerful PC Disk Drive, Confirms Plans to Join Two Divisions," *Wall Street Journal*, November 10, 1997.

32. Barnaby J. Feder, "Eureka! Labs with Profits," *New York Times*, September 9, 2001, B1; Robert Buderi, *Engines of Tomorrow: How the World's Best Companies Are Using Their Research Labs to Win the Future* (New York: Simon & Schuster, 2000).

33. George M. Whitesides, John P. Mathias, and Christopher Seto, "Molecular Self-Assembly and Nanochemistry: A Chemical Strategy for the Synthesis of Nanostructures," *Science* 1991, 254, 5036: 1312–19.

34. From "Editorial," by British scientist David Whitehouse, and "Subject Coverage," first issue of *Nanotechnology*, July 1990.

35. Michael L. Grienesien, "The Proliferation of Nano Journals," *Nature Nanotechnology* 2010, 5, 12: 825.

36. Andrew Pollack, "Atom by Atom, Scientists Build 'Invisible' Machines of the Future," *New York Times*, November 26, 1991.

37. Geof Bowker, "How to Be Universal: Some Cybernetic Strategies, 1943–1970," *Social Studies of Science* 1993, 23, 1: 107–27.

38. Even as late as 2006, leading researchers were still being asked to define nanotechnology. Editors, "nan´o·tech·nol´o·gy *n*.," *Nature Nanotechnology* 2006, 1, 1: 8–10.

39. See, for example, Ivan Amato, "The Apostle of Nanotechnology," *Science* 1991, 254, 6036: 1310–11; Robert Langreth, "Molecular Marvels," *Popular Science*, May 1993, 91–94, 110–11; David E. H. Jones, "Technical Boundless Optimism," *Nature* 1995, 374, 6525: 835–37; Gary Stix, "Waiting for Breakthroughs," *Scientific American* 1996, 274, 4: 94–99; David Voss, "Moses of the Nanoworld," *Technology Review* 1999, 60–62; David Rotman, "Will the Real Nanotech Please Stand Up?" *Technology Review*, March/April 1999, 46–53.

40. Phil Barth, "Nanoism: A new mass movement ideology," posted January 9, 1996, at sci.nanotech, http://groups.google.com/group/sci.nanotech/browse_frm/month/1996-01 (accessed March 2011), copy of posting in author's files.

41. Amato, "The Apostle of Nanotechnology," 1311.

42. Rotman, "Will the Real Nanotech Please Stand Up?" 53. Also, Eigler as quoted in Chris Toumey, "Apostolic Succession," *Engineering & Science* 2005, 1/2: 16–23.

43. K. Eric Drexler, "A tale of two nanotechnologies (a personal account of a peculiar history)," unpublished manuscript dated February 23, 2005, copy in author's possession.

44. Ed Regis, *Nano! Remaking the World Atom by Atom* (Boston: Little, Brown, 1995), 271. Charles Platt, "Nanotech: Engines of Hyperbole?" *Wired*, December 1993, 85–87.

45. K. Eric Drexler, "Letter to the Editor," *Science* 1992, 255, 5052: 268–69.

46. Stix, "Waiting for Breakthroughs," 94–99.

47. Ibid., 97 and 98.

48. Richard P. Feynman, "Cargo Cult Science," *Engineering and Science*, June 1974, 10–13.

49. Various letters in response to the *Scientific American* article as well as an in-depth discussion of it are at http://www.foresight.org/SciAmDebate/SciAm Letters.html#anchor378653 (accessed March 2011).

50. Phil Barth, January 17, 1996, posting to sci.nanotech, copy of posting in author's files.

51. For example, Lawrence Principe and William R. Newman, *Alchemy Tried in the Fire: Starkey, Boyle, and the Fate of Helmontian Chymistry* (Chicago: University of Chicago Press, 2002).

52. I am indebted to Michael Gordin for his extensive thoughts on this as well as sharing early drafts of his book *The Pseudoscience Wars: Immanuel Velikovsky and the Birth of the Modern Fringe* (Chicago: University of Chicago Press, 2012).

53. Rae Goodell, *The Visible Scientists* (New York: Little, Brown, 1977); also Jane Gregory and Steve Miller, *Science in Public: Communication, Culture, and Credibility* (New York: Plenum Publishing, 1998).

54. Bruce V. Lewenstein, "Cold Fusion and Hot History," *Osiris* 1992, 7, 135–63.

55. Stix, "Waiting for Breakthroughs," 98.

56. Mary Eisenhart, "Nanotechnology: Separating Myth from Reality (an Interview with Eric Drexler)," *Microtimes*, October 26, 1992, 182.

57. "New Technologies for a Sustainable World," June 26, 1992, hearing (S. HRG. 102-967) before the Subcommittee on Science, Technology, and Space of the Committee on Commerce, Science, and Transportation, United States Senate, 102nd Congress, pp. 22 and 25.

58. K. Eric Drexler, January 30, 2010, e-mail to the author.

59. Interagency Working Group on Nanoscience, Engineering, and Technology, *National Nanotechnology Initiative: Leading to the Next Industrial Revolution*, Washington, DC, 2000.

60. James S. Murday, May 29, 2007, oral history interview with Cyrus Mody, CHF.

61. Original announcement archived at http://www.nsf.gov/pubs/1998/nsf9820/nsf9820.txt (accessed March 2011).

62. M. C. Roco, "The US National Nanotechnology Initiative after 3 Years (2001–2003)," *Journal of Nanoparticle Research* 2004, 6: 1–10.

63. From April 22, 1998, testimony Lane gave to Congress; this was formerly available at Web site for the House Committee on Science (http://www.house.gov/science/), copy in author's possession. More Clinton-era science policy documents are at http://clinton3.nara.gov/WH/EOP/OSTP/html/OSTP_docarchives.html.

64. Thomas Kalil, June 12, 2006, interview with the author.

65. For example, M. C. Roco, "Government Nanotechnology Funding: An International Outlook," *JOM*, September 2002, 22–23.

66. Thomas Kalil, June 12, 2006, interview with the author.

67. Patrick Windham, "TPI Working Paper: The U.S. National Nanotechnology Initiative," Washington, DC, Technology Policy International, 2001. My thanks to Mr. Kalil for providing me with a copy of this white paper. Also, Neal Lane and Thomas Kalil, "The National Nanotechnology Initiative: Present at the Creation," *Issues in Science and Technology*, Summer 2005, 49–54.

68. M. C. Roco, S. Williams, and P. Alivisatos, eds., *Nanotechnology Research Directions: IWGN Workshop Report* (Baltimore: World Technology Evaluation Center, 1999), 179.

69. The phrase appears in a December 5, 1996, speech Greenspan, the former Federal Reserve chairman, made to the American Enterprise Institute.

70. Amato, *Nanotechnology*.

71. Alfred Nordmann, "Nanotechnology's Worldview: New Space for Old Cosmologies," *IEEE Technology and Society Magazine*, Winter 2004, 48–54.

72. Quotes and testimony from House of Representatives, Committee on Science, Subcommittee on Basic Research, *Nanotechnology: The State of Nano-Science and its Prospects for the Next Decade, 106th Congress* (Washington, DC: U.S. Government Printing Office, 1999).

73. Ibid.

74. Thomas Kalil, June 12, 2006, interview with the author.

75. William J. Clinton, January 21, 2000, speech, http://pr.caltech.edu/events/presidential_speech/pspeechtxt.html; also White House, Office of the Press Secretary, "National Nanotechnology Initiative: Leading to the Next Industrial Revolution," January 21, 2000, press release, http://clinton4.nara.gov/WH/New/html/20000121_4.html (accessed July 2010); John Markoff, "A Clinton Initiative in a Science of Smallness," *New York Times*, January 21, 2000, C5.

76. Quote appears in a number of places including M. C. Roco, "The US National Nanotechnology Initiative after 3 Years (2001–2003)," *Journal of Nanoparticle Research* 2004, 6: 1–10.

77. "Bucky Balls, Fullerenes, and the Future."

78. Vicki L. Colvin in her April 9, 2003, testimony, from "The Societal Implications of Nanotechnology: Hearing before the Committee on Science, House of Representatives (Washington, DC, 2003). This was previously available at http://www.house.gov/science/hearings/full03/apr09/colvin.htm; also, Kristen Kulinowski, "Nanotechnology: From 'Wow' to 'Yuck'?" *Bulletin of Science, Technology, and Society* 2004, 24, 1: 13–20.

79. Ronald Sandler, "The GMO-Nanotech (Dis)Analogy?" *Bulletin of Science, Technology, and Society* 2006, 26, 1: 57–62; Arie Rip, "Folk Theories of Nanotechnologists," *Science as Culture* 2006, 15, 4: 349–65.

80. Michael Powell, "Raving Robots, Mad Microbes: Tekkie Says Beware," *Washington Post*, April 16, 2000, F1.

81. Brent Schlender, "The Edison of the Internet," *Fortune*, February 1999, http://money.cnn.com/magazines/fortune/fortune_archive/1999/02/15/254898/index.htm (accessed April 2011).

82. Bill Joy, "Why the Future Doesn't Need Us," *Wired*, April 2000, 238–62.

83. Powell, "Raving Robots."

84. Joy, "Why the Future Doesn't Need Us."

85. For example, Edward Rothstein, "Even Techies Are Getting Nervous About Technology," *New York Times*, March 18, 2000, B9; Also, John Markoff, "Technologists Get a Warning and a Plea from One of Their Own," *New York Times*, March 13, 2000, C1.

86. Virginia Postrel, "Joy, to the World," *Reason*, June 2000, 46–48.

87. This image won the Vision of Science Award in 2002, which was sponsored by *Daily Telegraph* and the drug company Novartis. Brigitte Nerlich, "Powered by Imagination: Nanobots at the Science Photo Library," *Science as Culture* 2008, 17, 3: 269–92; also, Colin Milburn, *Nanovision: Engineering the Future* (Durham, NC: Duke University Press, 2008).

88. Adriana de Souza e Silva, "The Invisible Imaginary: Museum Spaces, Hybrid Reality and Nanotechnology," in *Nanoculture: Implications of the New Technoscience*, edited by N. Katherine Hayles (London: Intellect, 2004), 45.

89. July 11, 2000, e-mail from Roco to Smalley, folder 3, box 35, RS/CHF.

90. May 5, 2000, e-mail from Smalley to Richard O'Neill, folder 3, box 34, RS/CHF.

91. June 28, 2000, letter from Daniel S. Goldin to Smalley, folder 16, box 34, RS/CHF.

92. July 11, 2000, e-mail from Smalley to Roco, folder 3, box 35, RS/CHF.

93. Richard E. Smalley, "Of Chemistry, Love and Nanobots," *Scientific American*, September 2001, 76–77.

94. Bill Goldstein, "Honing the Science of the Release Date," *New York Times*, November 11, 2002, C9.

95. The ETC Group, *The Big Down: From Genomes to Atoms* (Winnipeg, Canada: ETC Group, 2003).

96. An example of a headline from spring 2003; this was from an Edinburgh weekly, http://scotlandonsunday.scotsman.com/uk/Charles-fears-science-could-kill.2422335.jp (accessed April 2011).

97. Martin Rees, *Our Final Century: Will the Human Race Survive the Twenty-first Century?* (New York: Random House, 2002); for Rees's bet, http://longbets.org/9/ (accessed April 2011).

98. Dennis Overbye, "It Was Fun While It Lasted," *New York Times Book Review*, May 18, 2003, 13.

99. Howard Lovy, "Nanotechnology Has Reached a Crossroads," *smalltimes*, July/August 2003, 6.

100. Michael Cobb and Jane Macoubrie, "Public Perceptions about Nano-technology: Risks, Benefits, and Trust," *Journal of Nanoparticle Research* 2004, 6, 4: 395–405.

101. Committee for the Review of the National Nanotechnology Initiative, *Small Wonders, Endless Frontiers: A Review of the National Nanotechnology Initiative* (Washington, DC: National Academy Press, 2003), 31.

102. Lovy, "Nanotechnology Has Reached a Crossroads," 6.

103. "Nanotechnology: Drexler and Smalley Make the Case for and against 'Molecular Assemblers,' " *Chemical & Engineering News* 2003, 81, 48: 37–42. There are several accounts of the Drexler-Smalley debate; I've drawn from Otávio Bueno, "The Drexler-Smalley Debate on Nanotechnology: Incommensurability at Work?" *HYLE—International Journal for the Philosophy of Chemistry* 2004, 10, 2: 83–98; David Berube and J. D. Shipman, "Denialism: Drexler vs. Roco," *IEEE Technology and Society Magazine* 2004, 23, 4: 22–26; and, especially, a fine article by Sarah Kaplan and Joanna Radin, "Bounding an Emerging Technology: Para-Scientific Media and the Drexler-Smalley Debate about Nanotechnology," *Social Studies of Science* 2011, 41, 4: 1–29.

104. Described in the table of contents for the *C&EN* issue.

105. Kenneth Chang, "Yes, They Can! No, They Can't: Charges Fly in Nano-bot Debate," *New York Times*, December 9, 2003, 2003, F3.

106. Quotes from "Nanotechnology: Drexler and Smalley."

107. Ibid.

108. Lawrence Lessig, "Stamping out Good Science," *Wired*, July 2004, http://www.wired.com/wired/archive/12.07/view.html?pg=5; Raymond Kurzweil, "The

Drexler-Smalley Debate on Molecular Assembly," December 1, 2003, http://www
.kurzweilai.net/the-drexler-smalley-debate-on-molecular-assembly (both accessed
April 2011).

109. K. Eric Drexler, December 12, 2010, e-mail to the author.

110. Howard Lovy, "Business Has Redefined Nano in Its Own Image," *smalltimes*, January/February 2004, 14.

111. Mark Modzelewski, "Nanotech Industry Can Help Groundbreaking Bill
Fulfill Its Promise," *smalltimes*, January/February 2004, 12.

112. Philip Ball, "Nanotechnology in the Firing Line," December 23, 2003,
online essay, http://nanotechweb.org/cws/article/indepth/18804 (accessed April
2011).

113. By 2011, chips with 22-nanometer-scale features were on engineers'
drawing boards.

Chapter 8: Visioneering's Value

1. Ed Regis. "The Incredible Shrinking Man." *Wired*, October 2004. 178–81,
204–5.

2. "Nanotechnology's Unhappy Father." *Economist*, March 11, 2004. http://
www.economist.com/node/2477051 (accessed April 2011).

3. Corie Lok, "Small Wonders," *Nature* 2010, 467, 7313: 18–21.

4. Drexler, quoted in "Nanotech Takes Small Step toward Buying 'Grey Goo,' "
Nature 2004, 429, 6992: 591.

5. K. Eric Drexler, "Nanotechnology: From Feynman to Funding," *Bulletin of
Science, Technology, and Society* 2004, 24, 1: 21–27.

6. K. Eric Drexler, "Productive Nanosystems," presentation given February 10,
2005, at National Academy of Sciences. FI.

7. Committee to Review the National Nanotechnology Initiative, *A Matter of
Size: Triennial Review of the National Nanotechnology Initiative* (Washington,
DC: National Research Council, 2006), 107.

8. January 8, 2010, e-mail from Drexler to the author.

9. Smalley describes this in "Nanotechnology: Drexler and Smalley Make the
Case for and against 'Molecular Assemblers,' " *Chemical and Engineering News*
2003, 81, 48: 37–42.

10. David Malakoff, "Congress Wants Studies of Nanotech's 'Dark Side,' " *Science* 2003, 301, 5629: 27.

11. Howard E. McCurdy, *Space and the American Imagination* (Washington,
DC: Smithsonian Institution Press, 1997).

12. John Schwartz, "Manned Private Craft Reaches Space in a Milestone for
Flight," *New York Times*, June 22, 2004, A1.

13. James Brooke, "Ambitious Entrepreneurs Planning to Send Tourists into
'Astronaut Altitude,' " *New York Times*, February 17, 1998, A16.

14. Chris Dubbs and Emeline Paat-Dahlstrom, *Realizing Tomorrow: The Path
to Private Spaceflight* (Lincoln: University of Nebraska Press, 2011).

15. Robin Snelson, "Unsung Heroes of the Personal Spaceflight Revolution."

Space Review 2004, published online September 27, 2004, http://www.thespace review.com/article/234/1 (accessed April 2011).

16. Observed on the office wall of sci-fi writer Jerry Pournelle when I interviewed him in 2007.

17. Rick Tumlinson, founder of the Space Frontier Foundation, quoted in John Schwartz, "Thrillionaires: The New Space Capitalists," *New York Times*, June 14, 2005, F1.

18. Nadrian C. Seeman, "Nanotechnology and the Double Helix," *Scientific American* 2004, 290, 6: 65–75.

19. Leonard M. Adleman, "Molecular Computation of Solutions to Combinatorial Problems," *Science* 1994, 266, 5187: 1021–24.

20. Paul W. K. Rothemund, "Folding DNA to Create Nanoscale Shapes and Patterns," *Nature* 2006, 440, 7082: 297–302.

21. The Kavli Prizes—$1,000,000 each—are awarded annually as a cooperative venture of the Norwegian Academy of Science and Letters, the Norwegian Ministry of Education and Research, and the Kavli Foundation; they are given in three categories: astrophysics, neuroscience, and nanoscience.

22. As described by Lewis Page in "Boffins Demo One-Molecule DNA 'Walker' Nano-bot: Tiny Four-Tentacle Machine Follows Origami Breadcrumbs," *The Register*, May 14, 2010, http://www.theregister.co.uk/2010/05/14/nano _dna_walker_robot/ (accessed April 2012).

23. The "assembly line" phrase comes from Hongzhou Gu et al., "A Proximity-Based Programmable DNA Nanoscale Assembly Line," *Nature* 2010, 465, 7295: 202–6. Both projects are described in Lloyd M. Smith, "Molecular Robots on the Move," *Nature* 2010, 465, 7295: 167–68; also, Gwyenth Dickey, "DNA on the Move," *Science News*, September 11, 2010, 18–21.

24. From http://news.uchicago.edu/article/2011/03/07/matthew-tirrell-named-founding-director-institute-molecular-engineering (accessed July 2011).

25. Esther Dyson et al., "Cyberspace and the American Dream: A Magna Carta for the Knowledge Age," in *Release 1.2*, 1994, http://www.pff.org/issues -pubs/futureinsights/fi1.2magnacarta.html (accessed April 2011).

26. A succinct introduction is Nick Bostrom, "A History of Transhumanist Thought," *Journal of Evolution and Technology* 2005, 14, 1: 1–27.

27. Julian Huxley, *Religion without Revelation* (London: Harper & Brothers, 1967 [1927]), 195.

28. Ed Regis, "Meet the Extropians," *Wired*, October 1994, http://www .wired.com/wired/archive/2.10/extropians.html (accessed June 2010).

29. Manfred E. Clynes and Nathan S. Kline, "Cyborgs and Space," *Astronautics* 1960, 26–27, 74–75; "Spaceman Is Seen as Man-Machine," *New York Times*, May 22, 1960, 31.

30. Quotes from *Extropy* #1, Fall 1988, copy in author's possession.

31. Background on Mondo 2000 comes from two main sources; a Web-based history under construction in early 2011 (http://www.mondo2000history.com/ history/) and a 1995 article in the *SF Weekly* by Jack Boulware called "Mondo 1995: Up and Down with the Next Millennium's First Magazine;" see http:// www.suck.com/daily/95/11/07/mondo1995.html (both accessed December 2010).

32. Rudy Rucker, R .U. Sirius, and Queen Mu, eds., *Mondo 2000: A User's Guide to the New Edge* (New York: HarperPerennial, 1992), 116.

33. From a 1989 "editorial" by R. U. Sirius and Queen Mu, quoted in Vivian Sobchack, "New Age Mutant Ninja Hackers: Reading *Mondo 2000*," in *Flame Wars: The Discourse of Cyberspace*, edited by Mark Dery (Durham, NC: Duke University Press, 1995), 11–28.

34. James Hughes, *Citizen Cyborg: Why Democratic Societies Must Respond to the Redesigned Human of the Future* (Cambridge, MA: Westview Press, 2004); Nicholas Agar, *Humanity's End: Why We Should Reject Radical Enhancement* (Cambridge, MA: MIT Press, 2010).

35. Vernor Vinge, "First Word," *Omni*, January 1983, 10. Vinge later acknowledged that the term originated, so far as he knew, with a tribute by mathematician Stanislaw Ulam to John von Neumann.

36. The best examples are his two books: *The Age of Spiritual Machines: When Computers Exceed Human Intelligence* (New York: Penguin Books, 1999) and *The Singularity Is Near: When Humans Transcend Biology* (New York: Viking, 2005).

37. Jeanne Clare Feron, "Machine Opens New Worlds to Blind," *New York Times*, April 13, 1980.

38. Raymond Kurzweil, "Live Forever," *Psychology Today*, January/February 2000, http://www.psychologytoday.com/articles/200001/live-forever (accessed April 2011).

39. James Gleick, "How Google Dominates Us," *New York Review of Books*, August 18, 2011, http://www.nybooks.com/articles/archives/2011/aug/18/how-google-dominates-us/ (accessed August 2011).

40. See http://www.kurzweilai.net/the-law-of-accelerating-returns (accessed August 2010).

41. Quote from Susan Hassler, "Un-assuming the Singularity," *IEEE Spectrum*, June 2008, 9.

42. From the Web site of the Singularity University (http://singularityu.org/about/faq/); also Jeffrey R. Young, "Will Electric Professors Dream of Virtual Tenure," *Chronicle of Higher Education*, November 28, 2008, http://chronicle.com/free/v55/i14/14a01301.htm, and Jeffrey R. Young, "What Traditional Academics Can Learn from a Futurist's University," *Chronicle of Higher Education*, September 14 2009, http://chronicle.com/article/What-Traditional-Scholars-Can/48369/ (all accessed May 2011).

43. Andrew Orlowski quoted in Ashlee Vance, "Merely Human? That's So Yesterday," *New York Times*, June 11, 2010, B1.

44. Francis Fukuyama, *Our Posthuman Future: Consequences of the Biotechnology Revolution* (New York: Farrar, Strauss & Giroux, 2002); Leon Kass, "*Preventing a Brave New World*: Why We Should Ban Human Cloning Now," *New Republic*, May 21, 2001, 30–39. Also see Jonathan Moreno, *The Body Politic: The Battle Over Science in America* (New York: Bellevue Literary Press, 2011).

45. A phrase used in the introduction of a special issue of *IEEE Spectrum*, from June 2008, devoted to Singularity-oriented ideas; also John M. Bozeman, "Technological Millenarianism in the United States," in *Millennium, Messiahs, and Mayhem: Contemporary Apocalyptic Movements*, edited by Thomas Rob-

bins and Susan J. Palmer (New York: Routledge, 1997), 139–58; Robert M. Geraci, *Apocalyptic AI: Visions of Heaven in Robotics, Artificial Intelligence, and Virtual Reality* (New York: Oxford University Press, 2010); David F. Noble, *The Religion of Technology: The Divinity of Man and the Spirit of Invention* (New York: Penguin Books, 1997).

46. Anna Lee Saxenian, *Regional Advantage: Culture and Competition in Silicon Valley and Route 128* (Cambridge, MA: MIT Press, 1994).

47. Ruth Oldenziel, *Making Technology Masculine: Men, Women, and Modern Machines in America, 1870–1945* (Amsterdam: Amsterdam University Press, 1999).

48. Joseph J. Corn, "Epilogue," in *Imagining Tomorrow: History, Technology, and the American Future*, edited by Joseph J. Corn (Cambridge: MIT Press, 1986), 219–29.

49. Gilbert Chin, Tara Marathe, and Leslie Roberts, "Doom or Vroom," *Science* 2011, 333, 6042: 538–39.

50. Frank Bruni, "In an Earthbound Era, Heaven Has to Wait," *New York Times*, July 6, 2011, A15.

51. Paul Gilding, quoted in Thomas L. Friedman, "The Earth Is Full," *New York Times*, June 8, 2011.

52. Michael Spector, "A Life of Its Own," *New Yorker*, September 28, 2009, 56–65. Also, Luis Campos, "That Was the Synthetic Biology That Was," in *Synthetic Biology: The Technoscience and Its Societal Consequences*, edited by Markus Schmidt (London: Springer, 2009), 5–21.

53. Kara Platoni, "Assembly Required," *Stanford*, July/August 2009, 44–49.

54. "DARPA to Offer \$30 million to Jump-Start Cellular Factories," reported on *Science*'s website, June 29, 2011, http://news.sciencemag.org/sciencein-sider/2011/06/darpa-to-offer-30-million-to-jump.html?ref=ra (accessed June 2011).

55. From January 2007 ETC report, http://www.etcgroup.org/en/node/602 (accessed August 2011).

56. Luis Campos, "Next Generation Nano? Narratives of Synthetic Biology," March 30, 2011, talk at the University of California, Santa Barbara, copy in author's possession.

57. Stewart Brand, *Whole Earth Discipline: An Ecopragmatist Manifesto* (New York: Viking Adult, 2009), 1. The original 1968 version, of course, was "We are as gods and might as well get good at it."

58. James Rodger Fleming, *Fixing the Sky: The Checkered History of Weather and Climate Control* (New York: Columbia University Press, 2010).

59. Dennis Overbye, "Offering Funds, U.S. Agency Dreams of Sending Humans to the Stars," *New York Times*, August 17, 2011, A1. Also, see the project's Web site, http://www.100yss.org/about.html (accessed September 2011).

60. "The 100-Year Starship Study: Strategy Planning Workshop," January 2011, http://www.100yss.org/resourcedownloads.html (accessed September 2011).

Index